FUNCTIONS WITH THE TI-84 PLUS & TI-84 PLUS SE

This book is dedicated to my wife and partner Teresa (Teri), whose continuing support has made this publication possible.

Brendan Kelly

Brendan Kelly Publishing Inc.
2122 Highview Drive
Burlington, Ontario L7R 3X4

ISBN 1-895997-25-9

ATTENTION EDUCATIONAL ORGANIZATIONS

Quantity discounts are available on bulk purchases of this book for educational purposes or fund-raising. For information, please contact:

Brendan Kelly Publishing Inc.
2122 Highview Drive
Burlington, Ontario
L7R 3X4

Web Site: www.brendankellypublishing.com

Telephone: (905) 335-3359 Fax: (905) 335-5104

INTRODUCTION

Memo

WHY IS THE GRAPHING CALCULATOR IMPORTANT?

The widespread acceptance of the graphing calculator as an important tool for the teaching and learning of mathematics is now evolving into a compelling assertion of its necessity. Mathematicians and mathematics educators recognize that this new technology, though not a panacea, is an essential mathematical tool that is changing irrevocably the content and pedagogy of high school and college mathematics.

TO THE STUDENT

Your TI-84 Plus calculator and this book will provide for you an exciting opportunity to discover many of the important concepts in mathematics. You will trace the trajectory of the longest home run in baseball history and calculate its height and range. You will graph the orbits of various planets around the sun and trace some of the historical milestones in the devclopment of mathematics and science. We encourage you to invest time complcting the exercises and investigations, checking your solutions against those in the back of the book and reflecting upon the concepts you encounter. When you have completed the explorations in this way, you will be in command of many of the most important ideas in mathematics. In your journey toward that destination, we hope you enjoy the content, the cartoons, and the entire experience!

Memo

WHY IS THIS BOOK NEEDED?

It is not feasible to expect students to master all the modes and menus of the TI-84 Plus before applying it to the study of functions. On the contrary, it is important that instruction in functions and pre-calculus draw upon the calculator functions as needed, neglecting those aspects that are not fundamental to the conceptual development. Achieving this integration of machine and mathematics requires a sequence of carefully designed instructional activities that showcase a technology-conscious approach to each statistical concept. This book is designed to be such a resource.

TO THE TEACHER

We believe that the teaching of mathematics to the next generation is a calling of the highest order. As you incorporate the graphing calculator into your mathematics classes, you will be introducing your students to the leading edge of instructional innovation in mathematics. This book and the others in our TI-84 Plus series have been designed to parallel our books for the TI-83 Plus. This enables you to involve *all* your students in the same lesson whether they have the TI-83 Plus or TI-83 Plus SE. Furthermore, your students will not be confused by alternative instructions designed to serve distinctly different calculators. In this way, we intend to support your professional efforts and to facilitate your task. Please share with us your comments and suggestions for future publications.

TABLE OF CONTENTS

© 1997 by Sidney Harris

"But we just don't have the technology to carry it out."

PART I — THE FUNCTION CONCEPT

TABLE OF CONTENTS

PART II — SPECIAL FUNCTIONS AND RELATIONS

This book attempts to teach mathematics and the use of the TI-84 Plus calculator at the same time. To do this effectively, it has been necessary to simulate the calculator keys and fonts on the TI-84 Plus and to generate screen displays. Achieving these elements required early access to beta versions of the TI-84 Plus and TI Connect. We are indebted to Len Catleugh, Vince Delisi, and Tom Ferrio of *Texas Instruments* whose continuing help and support in providing early versions of software and hardware have made these publications possible. We thank also Sarah Scott of *Texas Instruments* for her help in providing photos of the TI-84 Plus calculators for use on the cover of this book.

We acknowledge our debt to David Bernklau of Stuyvesant High School in New York City for his painstaking editing of the manuscript in its earlier drafts. Without his help, this book would have contained many omissions, glitches and goops. As in our other books, we have hidden a few more for you to find, but we trust they have been cleverly disguised. A special thanks to my wife, Teri, who purged the manuscript of a myriad typos, inconsistencies and non-sequiturs.

For a resource to be truly effective in the classroom, it must be free of ambiguities. Instruction must be clear and activities must be sequenced so that students can learn easily from it. The only way to achieve this is to conduct extensive field testing. We are indebted to the students enrolled in the pre-service teaching program at OISE/UT (University of Toronto) for working through versions of earlier drafts of the manuscript and providing innumerable improvements. It is only through such field testing that we have been able to anticipate potential student difficulties and make the necessary modifications to facilitate your role as teacher.

Part I
The Evolution of The Function Concept

The initial impetus for the development of symbolic algebra was the quest for roots of equations. As mathematicians began to study motion, their search for the points at which a graph intersects the x-axis (quest for roots) expanded to the investigation of the shape and properties of the graph. For example, in the study of the motion of a projectile, mathematicians were interested in the shape of the trajectory, the maximum height attained and the range. This interest led to the development of the concept of a mathematical function. The highlights of this evolution showing the increasing generalization in the definition of a function are outlined below.

Leonhard Euler 1707-1783

This definition of function, attributed to Leonhard Euler, prevailed until 1807 when Jean Joseph Fourier described heat flow using a "function" defined differently for different intervals on the x-axis. It soon became clear that a new, more general definition of a function would be required to include such cases.

↓

•1734 *A function $f(x)$ is any algebraic expression involving variables and constants defined by an equation or graph.*

Georg Cantor 1845-1918

•1837 *If two variables, x and y are so related that whenever a value is assigned to x there is automatically assigned by some rule of correspondence, a value to y, then we say y is a "function" of x.*

Lejeune Dirichlet 1805-1859

A function, f, is any set of ordered pairs of elements such that if $(x_1, y_1) \in f$, $(x_2, y_2) \in f$ and $x_1 = x_2$, then $y_1 = y_2$.

The set of all first elements of the ordered pairs is called the "domain" of the function and the set of all second elements of the ordered pairs is called the "range" of the function.

The definition of a function as a correspondence was given by Dirichlet. He went on to define the permissible values of x as the *domain* of the function and the values assumed by y as the *range*. This definition removed the need to define y as an algebraic expression in x. The final step in generalizing the function concept was taken by Georg Cantor near the end of the 19th century. His definition generalized the function concept so that it is expressed in terms of elements of sets and is therefore independent of the concepts of number and variable.

EXPLORATION 1 Tables: The Earliest Functions

From earliest recorded history, the quest for understanding has led the human intellect to the investigation of mathematical ideas. Information about the ancient Egyptian mathematics has come to us from a few papyri (old "paper"scrolls). The most important of these is the *Rhind Mathematical Papyrus* found in the ruins of a small building at Thebes on the Nile. It was purchased in 1858 by Henry Rhind and donated to the British museum on his death.

This document was copied by the scribe, Ahmes, in about 1650 B.C. and described about 87 mathematical problems and their solutions that may have been developed as early as 2500 B.C.! The Rhind papyrus was written in *hieratic*, a kind of "italic" form of hieroglyphics. In the figure, you see a copy of Problem 79 the way it appears in the papyrus and below it you see the translation of the hieratic into hierglyphics and English.

The translation reveals that Problem 79 was the forerunner and perhaps the origin of the familiar related nursery rhyme:

As I was going to St. Ives,
I met a man with seven wives;
Every wife had seven sacks,
Every sack had seven cats,
Every cat had seven kits.
Kits, cats, sacks, and wives,
How many were going to St. Ives?

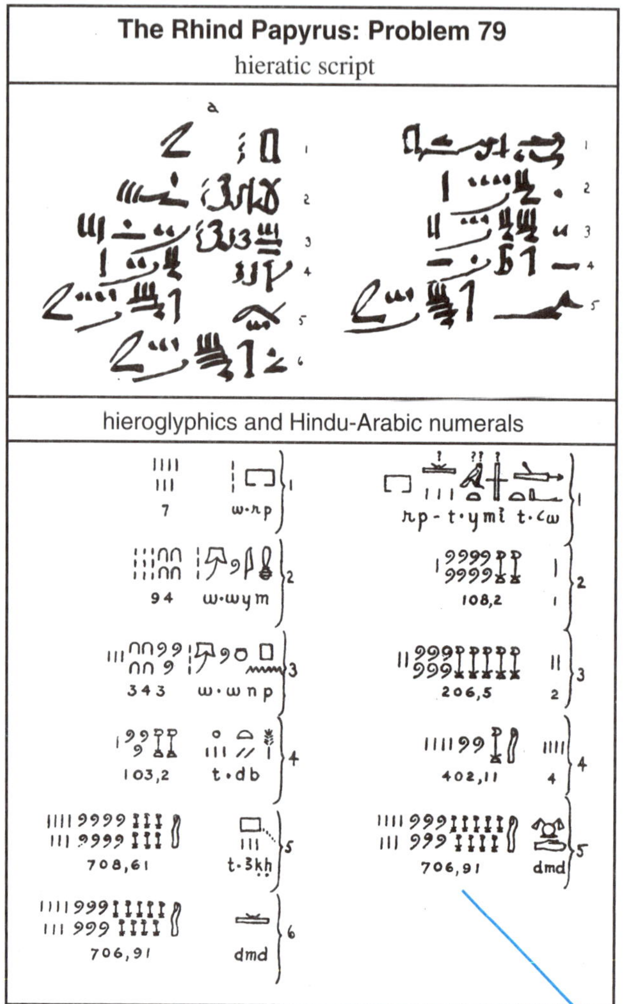

Problem 79 asks that we evaluate the sum, $7 + 7^2 + 7^3 + 7^4 + 7^5$. In the column on the left, Ahmes lists the powers of 7 obtained by multiplying each term by 7 to get the next term. He then writes the sum of these powers. On the right, he shows that we can evaluate $7 + 7^2 + 7^3 + 7^4 + 7^5$ by adding 1 to $7 + 7^2 + 7^3 + 7^4$ and multiplying this result (2801) by 7.

Observe that the Egyptian numbers are written in reverse so that 49 appears as 94 and 19,607 appears as 706,91.

The Rhind papyrus and the few other papyri which have survived show that the earliest relationships among numbers were displayed as tables. The development of tables was the first step toward the development of formulas and ultimately equations and functions.

WORKED EXAMPLE

a) Write a formula for the positive powers of 7 and for the sum of the first n positive powers of 7.

b) Create a table to display the first n positive powers of 7 and the sum of these powers up to the nth.

Solution

a) The positive powers of 7 can be written as 7^n where $n = 1, 2, 3, \ldots$ To evaluate $7^1 + 7^2 + 7^3 + \ldots + 7^n$, we let $S = 7^1 + 7^2 + 7^3 + \ldots + 7^n$. Then following the method in the Rhind papyrus, we add 1 to S and multiply by 7, to get the sum of the first $n + 1$ powers of 7.

$$7(1 + S) = 7(1 + 7^1 + 7^2 + 7^3 + \ldots + 7^n)$$

Expanding yields $$7 + 7S = 7^1 + 7^2 + 7^3 + \ldots + 7^n + 7^{n+1}$$

Subtracting S from both sides $$S = 7^1 + 7^2 + 7^3 + \ldots + 7^n$$

$$7 + 6S = 7^{n+1}$$

Solving for S yields, $S = \dfrac{7^{n+1} - 7}{6}$.

BEFORE YOU USE YOUR TI-84 Plus

Make sure your calculator is set in the default modes. Press [MODE]. Your screen should look like this:

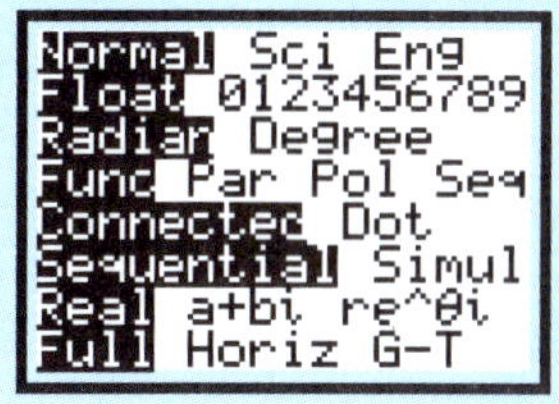

If one or more settings in the left column are not highlighted, press [▼] repeatedly until you reach the desired setting and press [ENTER].

b) To define $Y_1 = 7^n$ we press these keys: [Y=] [7] [^] [X,T,θ,n] [ENTER]

Note that the TI-84 Plus uses the variable X instead of n.

To define $Y_2 = \dfrac{7^{n+1} - 7}{6}$, we press these keys:

[Y=] [(] [7] [^] [(] [X,T,θ,n] [+] [1] [)] [−] [7] [)] [÷] [6] [ENTER]

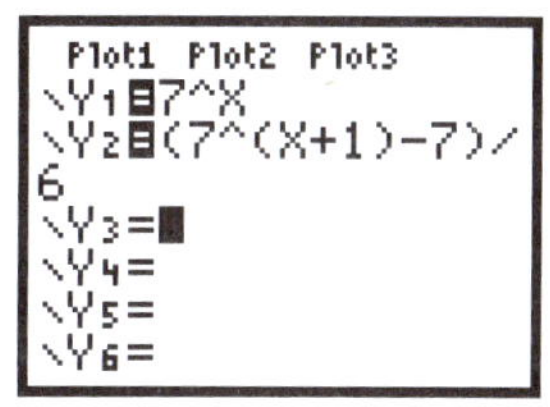

Before creating a table, we must identify the values of the variable, n (whoops, X) for which the table is displayed. To do this, we press: [2nd] [TBLSET] to obtain this display.

The cursor in the display is flashing over a zero setting. Since we want only positive values of the variable, we press:

[1] [ENTER] to change the setting to TblStart = 1.

The ΔTbl = 1 indicates that X will be incremented by 1 in successive rows.

To create a table, we press [2nd] [TABLE]. We obtain this display. The column under Y_1 shows the positive powers of 7 and the column under Y_2 shows the cumulative sums of these powers; i.e.

$$7, \quad 7 + 7^2, \quad 7 + 7^2 + 7^3$$

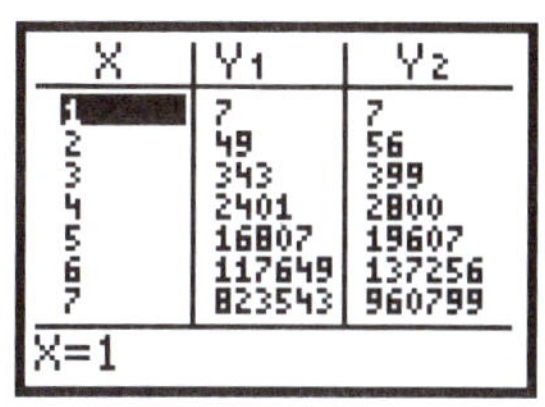

Compare the numbers in our table with the numbers in the Rhind papyrus (reading the digits in reverse order). We observe that they agree for the first four powers of 7, except for 7^4, which the scribe wrote as 2301 instead of 2401. It is significant that such questions were pondered over 4000 years ago by civilizations with rudimentary technologies. Today, we merely press [▼] to scroll down the table and peruse the powers of 7!

Note: If the display on your screen is difficult to see, you can adjust your screen contrast by repeated pressing of these keys:

- to increase contrast: [2nd] [▲]
- to decrease contrast: [2nd] [▼]

Exercises

1. Use the method in the *worked example* to find:
   a) a formula for the sum of the first n positive powers of 9.

   b) a formula for the sum of the geometric series, $a + ar + ar^2 + ar^3 + \ldots + ar^n$ where $r \neq 1$.

2. a) Use the formula you derived in *exercise* 1 to create a table showing the first n positive powers of 9 in one column and the sum of these powers up to the nth in the next column.

   b) Determine from your table the value of 9^6 and the value of $9 + 9^2 + 9^3 + 9^4 + 9^5 + 9^6$.

   c) For each power of 9 in your table find the sum of the digits. (e.g. The sum of the digits of 729 is $7 + 2 + 9 = 18$). Discuss any pattern you find. Is this pattern true for the numbers in both columns of your table?

3. Create a table to display 2^n for positive values of n. Scroll down in your table to find the smallest value of n such that.

   a) $2^n > 10^6$ b) $2^n > 10^9$ c) $2^n > 10^{12}$

4. According to an old legend, the inventor of chess made a strange request as his reward for his invention. He asked that a grain of wheat be placed on the first square of a chess board, 2 grains on the next square, 4 on the third, and so on, each time doubling the number of grains on the previous square. Make a table to help you answer these questions.

   a) What is the first square on which the number of grains of wheat exceeds 1 million? 1 billion?

   b) What is the first square for which the total number on all squares exceeds 1 trillion?

   c) What is the total number of grains on all 64 squares? Without using your table, calculate this number to 10 significant digits.

Exercises

5. a) Create a table that displays 7^{n+5} and 9^n for positive values of n. Scroll down in your table to find the value of n such that $9^{n-1} < 7^{n+5} < 9^n$.

   b) Solve the inequality in part a) by creating a table of values of Y_1 for positive values of n where

$$Y_1 = \frac{9^n}{7^{n+5}}$$

CHALLENGE

6. The diagram shows regular polygons of 4, 5, 6, and 7 sides and all their diagonals.

   a) Record the number of sides and diagonals of each.

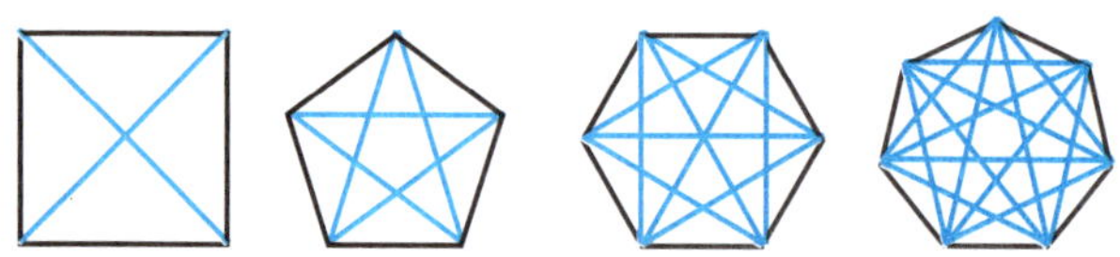

   b) Enter your numbers in a table like the one below.

   c) Study your table to find a relationship between the number of sides and the number of diagonals. Use this relationship to write an expression for the number of diagonals in a polygon with n sides.

Number of Sides	Number of Diagonals
4	2
5	5
6	9
•	•
•	•

   d) Construct a table on your TI-84 Plus that gives the number of diagonals corresponding to the number of sides of a polygon.

   e) Use the table you constructed in Part d) to determine the number of diagonals in a *dodecagon* (i.e. a polygon of 12 sides). Then use the diagram of the regular dodecagon shown below to check your answer.

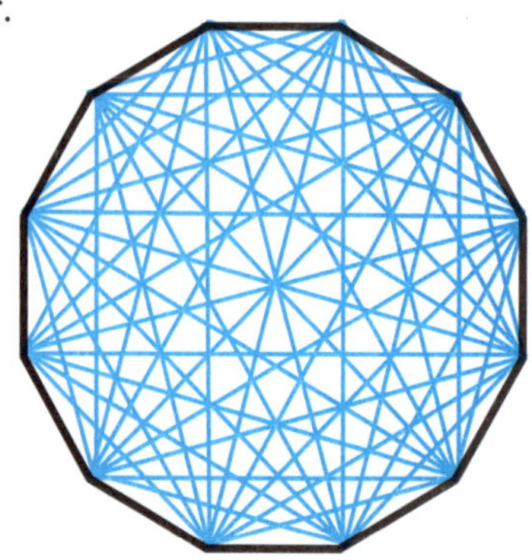

Investigation: Figurate Numbers

Pythagoras
c. 585 B.C.– 501 B.C.

In ancient Greece around 530 B.C. a group of philosophers and mathematicians formed a secret society dedicated to the study of numbers and their properties. This secret society was known as the *Pythagoreans*, in honor of their distinguished leader, Pythagoras.

Believing that the "personalities" of numbers determined the character of the objects they described, the Pythagoreans associated numbers with shapes. Numbers that can be represented by triangular arrays of dots (such that the nth row contains n dots) were called *triangular numbers*. Numbers that can be displayed in n rows of n dots were called *square numbers*, and so on.

The first four triangular numbers are shown here using dots.

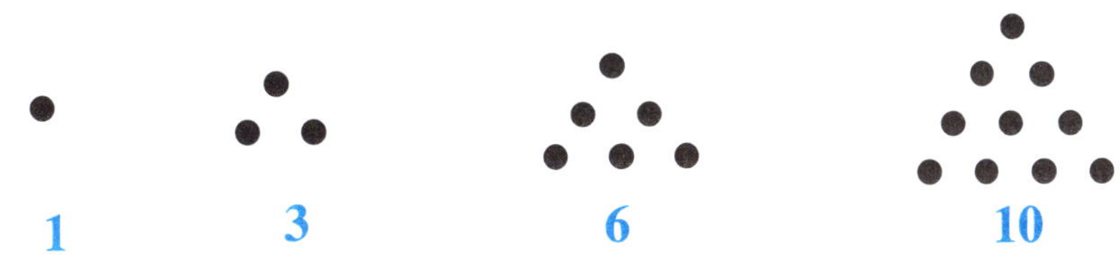

a) Complete the sequence to show the first ten triangular numbers.

1, 3, 6, 10, _, _, _, _, _, _

b) The diagram shows a rectangular array of n rows of dots containing $n + 1$ dots in each row. Write the number of dots in the array in terms of n.

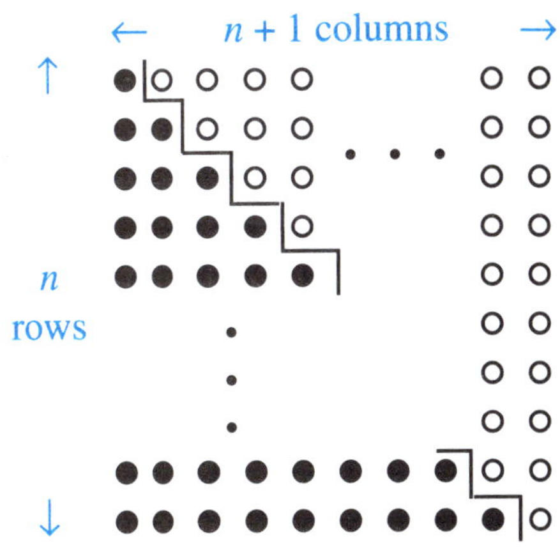

c) Write the number of black dots as a fraction of the total number of dots in the array.

d) Use the relationship you found in Part c) to write an algebraic expression for the number of black dots in the rectangular array.

e) Use the expression you found in Part d) to write an algebraic expression for the nth triangular number.

f) Write an algebraic expression for the sum of the integers from 1 to n, i.e., $1 + 2 + 3 + \ldots + n$.

g) Make a table (like the one shown here) in your notebook that lists for each integral value of n the number of diagonals in a polygon of n sides and the nth triangular number. Write a sentence to describe the relationship between the numbers in the second and third columns of your table.

Number of Sides	Number of Diagonals	Triangular Number
4	2	3
5	5	6
6	9	10
•	•	•
•	•	•

h) Use the relationship you discovered in Part g) to write an expression for the number of diagonals in a polygon of n sides.

i) On your TI-84 Plus, create a table like the one shown and use it to determine the number of diagonals in a polygon of 30 sides.

EXPLORATION 2 The Graph: Snapshot of a Function

Cogito ergo sum.
(I think, therefore I am.)
René Descartes 1596-1650

On November 10, 1619, a brilliant French mathematician and philosopher changed irrevocably the course of mathematical development by unifying algebra and geometry, the two main branches of mathematics. René Descartes reported that the inspiration for his discovery of analytic geometry flashed before him in a vivid dream. The essence of his idea was the representation of algebraic equations by sets of points in a plane. Today we refer to these sets of points as *graphs.* How remarkable it is that the concept of a graph, spawned four centuries ago in a moment of inspiration, has become a cornerstone of modern mathematics. Furthermore, the new technology renders the graph of an equation instantly with the push of a few buttons!

Before Descartes, functions were defined by tables of values (i.e. sets of ordered pairs) or by equations. Descartes' great idea was to display each ordered pair (x, y) as the location that is x units to the right of a vertical line called the *y-axis* and y units above a horizontal line called the *x-axis*. The set of all such ordered pairs defining a function are therefore displayed as a set of points in the plane called the *graph of the function*.

An ordered pair (x, y) is displayed as a point in the (first quadrant of the) x-y plane having coordinates x and y.

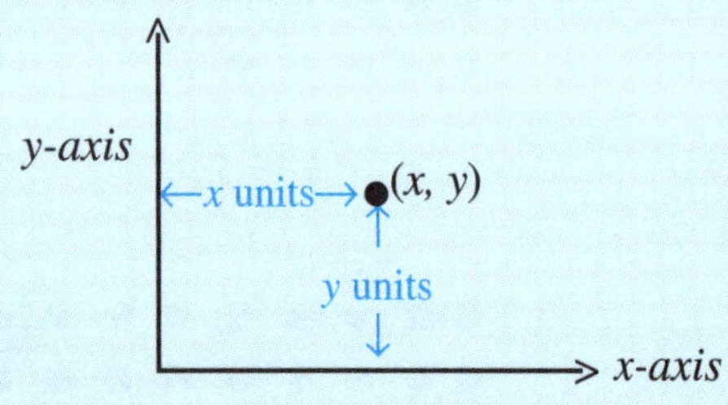

The x-y plane is divided into four sections or *quadrants* so that points left of the y-axis are assigned negative x-coordinates and points below the x-axis are assigned negative y-coordinates.

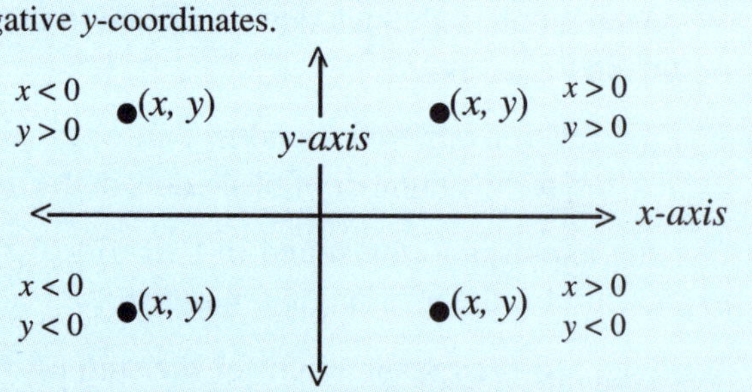

A graph of the set of triangular numbers given by: $n(n + 1)/2$ where $n = 1, 2, 3, \ldots, 10$

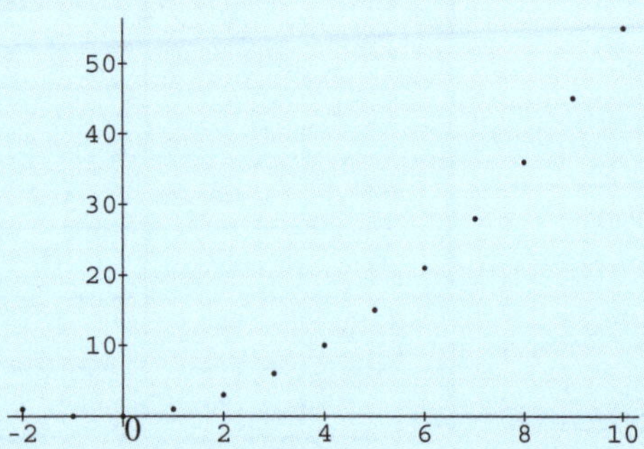

A graph of the set of numbers of the form $n(n + 1)/2$ where n is a whole number such that $-6 \leq n \leq 6$

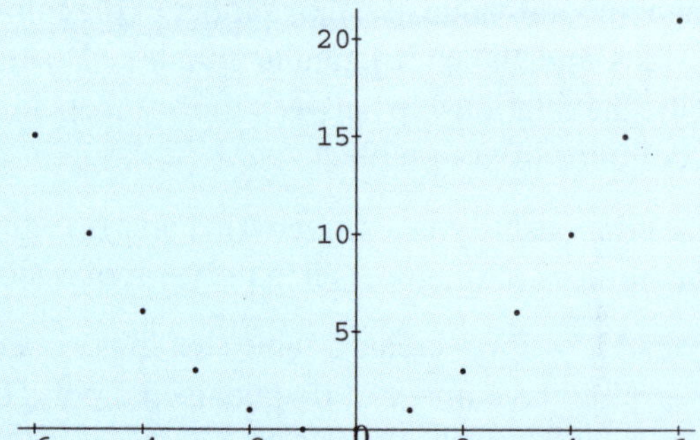

The worked examples that follow show how to construct the graphs of functions that are defined by equations.

WORKED EXAMPLE 1

a) Sketch the graph defined by the equation $u(n) = n(n + 1)/2$ for $n = 1, 2, 3, \ldots, 10$.
b) Trace along your graph to determine the value of $u(7)$
c) Determine the value of $u(100)$

Solution

a) Since n is a whole number, the equation defines the set of ordered pairs
$\{(1, 1), (2, 3), (3, 6), \ldots (n, n(n + 1)/2), \ldots, (10, 55)\}$

The values $u(n)$ form the sequence of triangular numbers. To graph such a sequence, we must select *sequence* mode by pressing these keys:

MODE ▼ ▼ ▼ ▶ ▶ ▶ ENTER

To enter the equation that defines the sequence of triangular numbers,

we press Y= .

To enter the defining equation $u(n) = n(n + 1)/2$, we press:

X,T,θ,n (X,T,θ,n + 1) ÷ 2 ENTER

and we obtain the display shown on the right.

To define the first term $u(1) = 1$, we press: 1 ENTER and the brackets { } are automatically placed around the 1 as shown in the display.

To graph the first 10 terms of $u(n)$, we press WINDOW and set the window variables as shown in the display.

When we press GRAPH , we obtain the graph shown in the display below left.

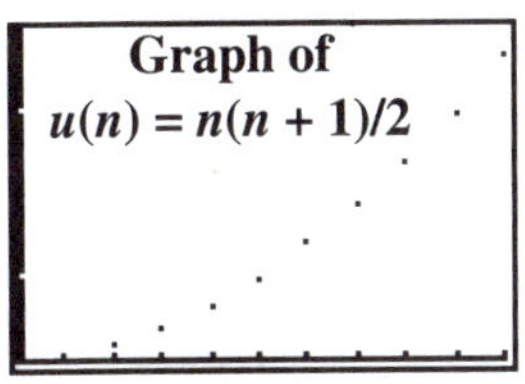

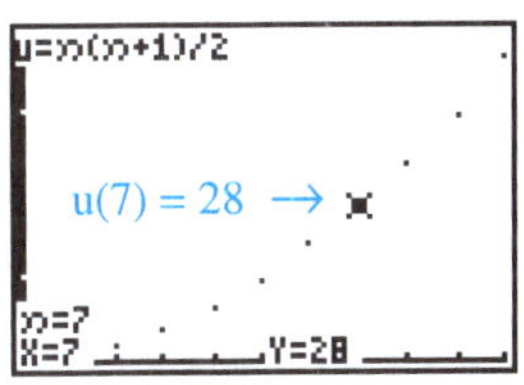

b) When we press TRACE and then repeatedly press the ▶ key, we move along the graph of the sequence as shown. The display shows that $u(7) = 28$.

c) To determine $u(100)$, press WINDOW and set nMax to 100 as in the display. Then press: GRAPH 2nd [CALC] ENTER to obtain the display below left. Then enter 100 to obtain the display below right indicating that $u(100) = 5050$.

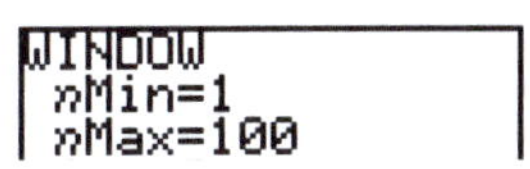

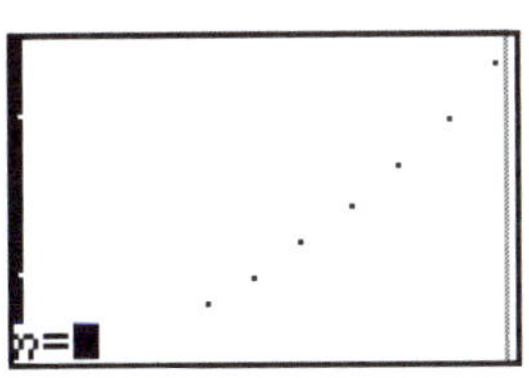

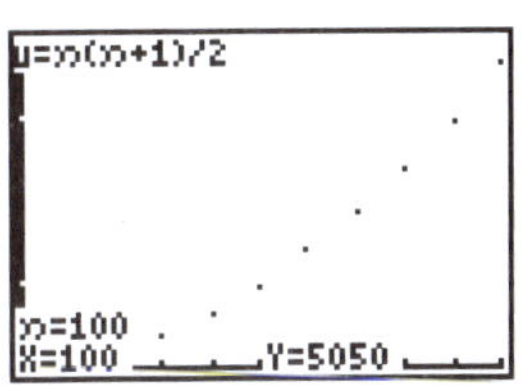

In *worked example* 1, we constructed the graph of the function $u(n)$ where n is a variable that takes only positive integral values. Such functions are called *sequences* and are graphed in *sequence* mode. Functions that are defined for any real number x are called functions of a real variable and are graphed in *function* mode. While the graphs of sequences are discrete (i.e. disconnected) sets of points, the graphs of functions of a real variable are usually (but not always) composed of one or more continuous curves. In the next example we graph such a function.

WORKED EXAMPLE 2

Sketch the graph defined by the equation $y = x(x + 1)/2$ in the interval $-10 \le x \le 10$ and display the graph for the range $0 \le y \le 60$. Find the value of y when $x = 7$.

Solution

Before we can graph the function defined by this equation, we must choose *function* mode by pressing these keys:

MODE ▼ ▼ ▼ ENTER

Function mode has been selected →

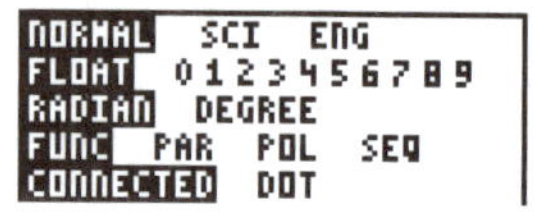

Then enter the expression $x(x + 1)/2$ by pressing:

Y = X,T,θ,*n* (X,T,θ,*n* + 1) ÷ 2 ENTER

Function mode is the default mode, so you will not usually need to choose function mode unless you've been working in another mode.

To display the graph of this function in the interval $-10 \le x \le 10$ and for the range $0 \le y \le 60$, we press WINDOW and set the window variables as shown on the right.

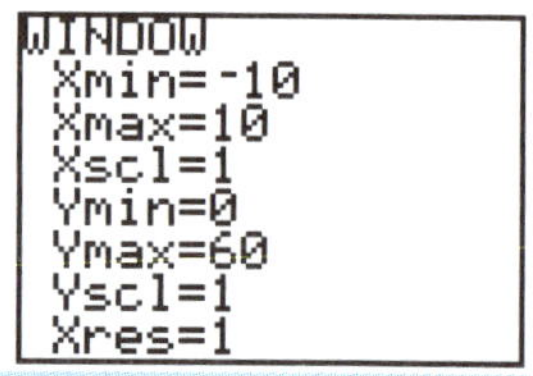

Upon entering GRAPH we obtain the graph shown below.

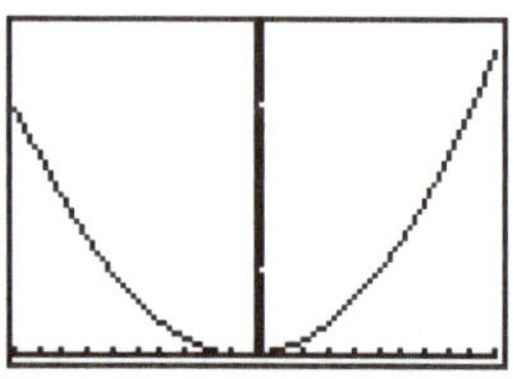

The default window settings are: $-10 \le x \le 10$ and $-10 \le y \le 10$ so it may be necessary to adjust only the y values

One way to obtain the value of y when $x = 7$ is to press TRACE and use the ▶ key to move along the curve to the point (7.02…, 28.15…) shown in the display. Since y is an integer when x is an integer, then y must be 28 when $x = 7$.

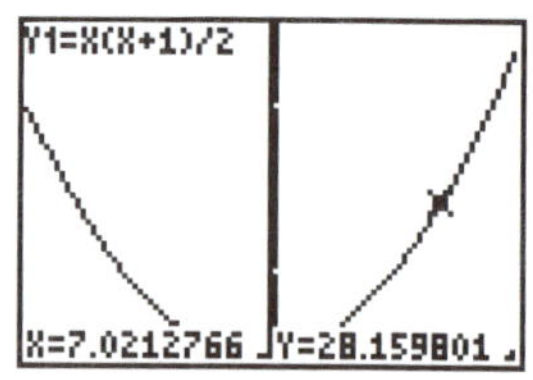

Alternatively, the exact value of y when $x = 7$ can be obtained by using the calculation menu accessed by pressing: 2nd [CALC] ENTER and entering 7 ENTER in response to the prompt for x. This yields the display on the right showing that $y = 28$ when $x = 7$.

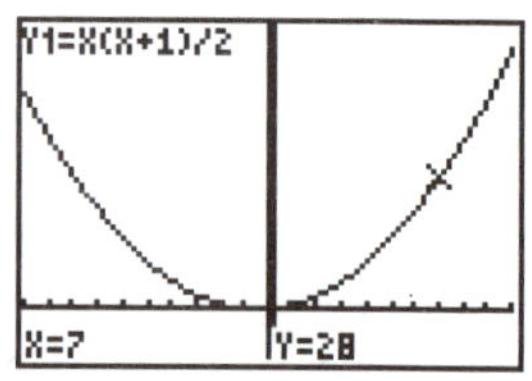

We could further verify this answer by accessing the table of values through the keying sequence 2nd [TABLE] and moving the cursor down to $x = 7$.

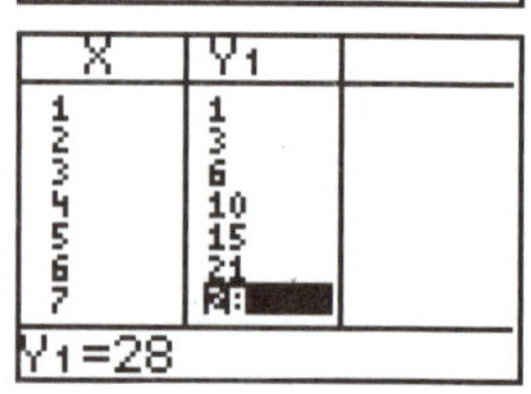

WORKED EXAMPLE 3

a) Sketch the graph defined by the equation $\frac{1}{8}x^3 + 2$ in the window $-10 \le x \le 10; -10 \le y \le 10$

b) Find the y-coordinate of the point on the graph with $x = -2$.

Solution

a) To graph the equation, $\frac{1}{8}x^3 + 2$, we press: [Y=]

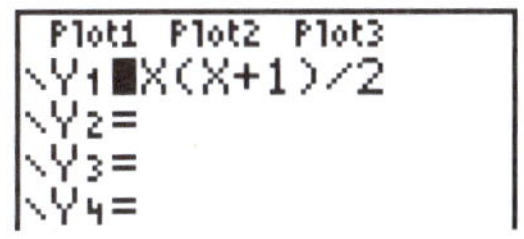

If you completed *worked example* 2 your calculator screen may look like this after you press the [Y=] key.

To position the cursor and define Y_2 we press these keys:

[▼] [X,T,θ,n] [^] [3] [÷] [8] [+] [2] [ENTER]

To obtain the default window, $-10 \le x \le 10; -10 \le y \le 10$, and graph the functions, we press [ZOOM] [6]

The graphs of Y_1 and Y_2 are shown in this display.

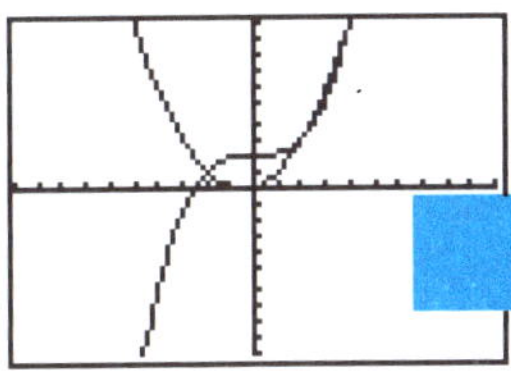

How do we graph only one equation at a time?

To graph only Y_2 and not Y_1, press [Y=] and use the cursor keys to move the cursor over the "=" sign in the equation defining Y_1. Press [ENTER] . This de-selects Y_1 and then you can graph Y_2 by pressing [GRAPH] .

The display shows the graph of Y_2 only.

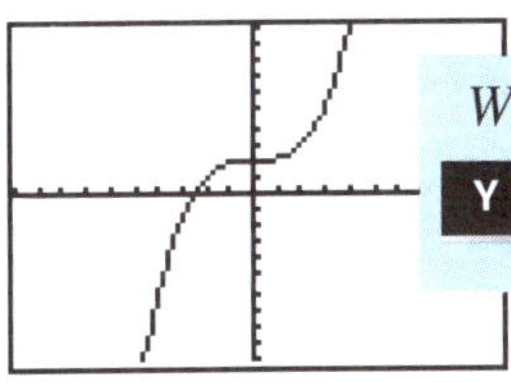

Would we obtain the graph of Y_2 using the following sequence? Why or why not?

[Y=] [1] [÷] [8] [X,T,θ,n] [^] [3] [+] [2] [ENTER] [GRAPH]

In the investigation section, you will learn how to use the [TRACE] key to approximate the x-intercepts of various graphs.

b) To determine the y-coordinate of the point on the graph with $x = -2$, we press: [2nd] [CALC] [ENTER]

This yields the display shown below left. In response to the prompt for a value of x, we key in [(−)] [2] [ENTER] and we obtain the display below right.

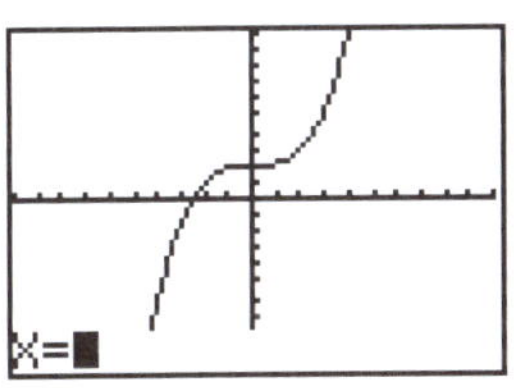

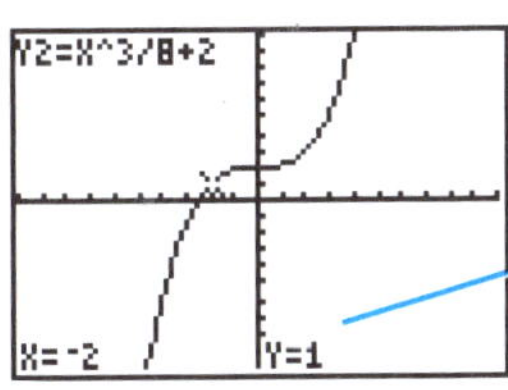

This display shows that 1 is the y-coordinate of the point on the graph with $x = -2$.

Exercises

1. The graph below shows the volume of water in Archimedes' bathtub during his 45-minute bath.

Use the graph to help you answer the following questions.

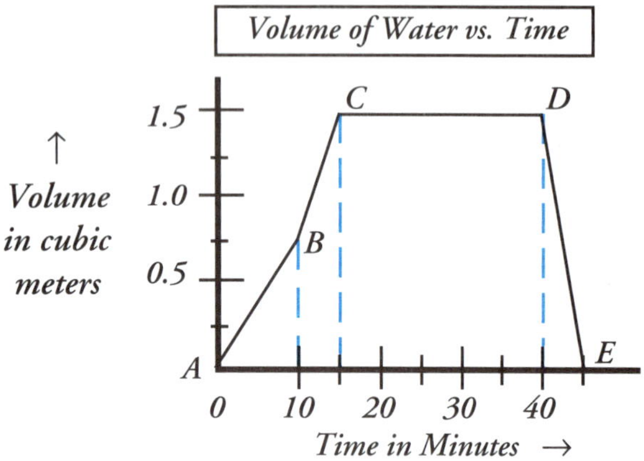

a) What is the volume of water in the tub after 10 minutes? 15 minutes? 40 minutes?

b) What is the slope of line segment CD? What does this tell us about the flow of water into the tub?

c) What are the slopes of line segments AB, BC and DE? Interpret each slope in terms of water flow.

d) Do you think that Archimedes had a hot water tap?

e) How would the graph differ from the one above if we graphed the *height* of the water in the tub against time?

2. Sketch the graph defined by the equation
$u(n) = n^2 - 3n$ for $n = 1, 2, 3, \ldots, 10$.
(Remember to access *sequence* mode.)

Press WINDOW and set the range of x values as shown in the display.

```
WINDOW
 nMin=1
 nMax=10
 PlotStart=1
 PlotStep=1
 Xmin=0
 Xmax=10
↓Xscl=1
```

Then press ZOOM 0 to have the calculator select the appropriate range for y and graph the equation. Trace to find the value $u(8)$.

3. Graph each function with the default window settings.

$$y = x^4 - 7x^2 - x + 6$$

and $$y = -x^4 + 7x^2 + x - 6$$

Note: To enter the $-x^4$ term, use the (–) key and not the subtraction symbol.

How are these two graphs related?

Investigations

4. Graph each of the following equations on your TI-84 Plus calculator. Press TRACE and use the arrow keys to trace along each graph to estimate the y-value corresponding to $x = -5$. In what way are the graphs e) and f) alike? Can you explain why?

a) $y = 3x - 7$ b) $y = 2x^2 - 4$
c) $y = 2x^3 - 6x^2 - x + 6$ d) $y = x^4 - 7x^2 - x + 6$
e) $y = x^5 - 3x^3 + 4$ f) $y = -x^5 + 3x^3 + 4$

5. Define on your TI-84 Plus calculator the function $Y_1 = x^4 - 75x^2 + 400$.

a) Graph Y_1 using the default settings; i.e., press ZOOM 6

b) Then fit the range to show the entire graph in the interval $-10 \le x \le 10$ by pressing ZOOM 0.

c) Use 2nd [CALC] ENTER to calculate the exact value of Y_1 corresponding to $x = -6$.

d) Describe any symmetry in this graph and explain how you could recognize this symmetry by looking at the equation. Deduce the value of Y_1 at $x = 6$.

e) Describe how changing the coefficient of the x^2 term changes the shape of the graph.

6. Define on your TI-84 Plus calculator the function $Y_1 = x^5 - 7x^4 - 17x^3 + 119x^2 + 16x - 112$.

a) Graph Y_1 using the default settings; i.e., press ZOOM 6

b) To redraw the graph using appropriate range settings for Ymin and Ymax, press ZOOM 0

c) To view the part of the graph close to the x-axis, draw a viewing box by following these instructions:
Press ZOOM ENTER.
Using the arrow keys, move the cursor to the point with coordinates (–5.53…, 6169…) and press ENTER.
This fixes the upper left corner of your viewing box. To fix the lower right corner, move the cursor to the point with coordinates (8.93…, –7242…) and press ENTER.

d) Repeat step c) to create another box that displays the graph in the region close to the x-axis. Record the approximate values of x at which the graph crosses the x-axis. Do you think there are any other values of x at which the graph crosses the x-axis? Explain.

EXPLORATION 3 The Linear Function & its Inverse

Anders Celsius
1701-1744

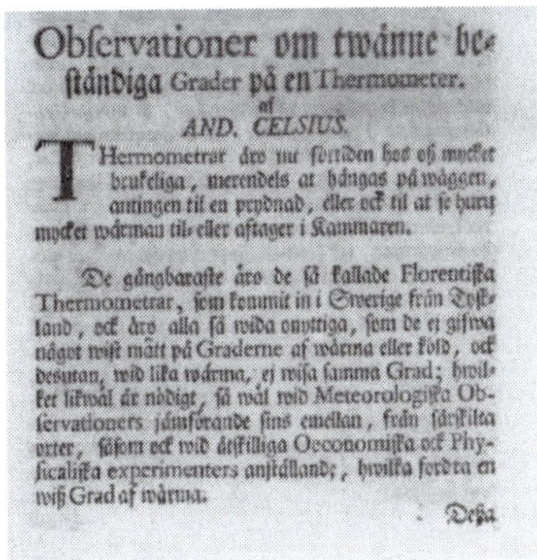

Obſervationer om twänne beſtåndiga Grader på en Thermometer.
af
AND. CELSIUS.

THermometrar äro nu förtiden hos oß mycket brukeliga, merendels at hängas på wäggen, antingen til en prydnad, eller ock til at ſe huru mycket wärman til- eller aftager i Kammaren.

De gångbaraſte äro de ſå kallade Florentiſka Thermometrar, ſom kommit in i Swerige från Tyſkland, ock äro alla ſå wida onyttiga, ſom de ej gifwa något wiſt mått på Graderne af wärma eller köld, ock deſutan, wid lika wärma, ej wiſa ſamma Grad; hwilket likwäl är nödigt, ſå wäl wid Meteorologiſka Obſervationers jämförande ſins emellan, från ſärſkilta orter, ſåſom ock wid åtſkilliga Oeconomiſka ock Phyſicaliſka experimenters anſtällande, hwilka fordra en wiß Grad af wärma.

Deßa

"LET'S GO OVER TO CELSIUS'S PLACE. I HEAR IT'S ONLY 36° OVER THERE."

In 1742 Swedish astronomer Anders Celsius published a paper titled, *Observations on Two Persistent Points on a Thermometer.* In that paper, Celsius proposed that the temperature scale be defined by two fixed points:

- the temperature at which water boils
- the temperature at which water freezes.

He originally proposed that the temperature at which water boils be assigned the value 0 and the temperature at which water freezes be assigned the value 100. A celsius degree would then be one-hundredth of the difference between these temperatures. After Celsius' death, his temperature scale was adopted but in a reversed form, i.e. the temperature of boiling water was fixed at 100° C and freezing water at 0°C. If the scale had not been reversed, the gentlemen in the cartoon would have found Celsius' place to be even more unbearable!

Before the Celsius temperature scale had been adopted, the world used the Fahrenheit scale named in honor of German physicist Gabriel Daniel Fahrenheit (1686–1736). On that scale, the freezing point of water is 32°F and the boiling point of water is 212°F.

The natural question that arises is: *What is the mathematical equation that converts a temperature in degrees Celsius to degrees Fahrenheit?* To answer this, we observe that there are 180 Fahrenheit degrees but only 100 Celsius degrees between the freezing and boiling points of water. So each Celsius degree corresponds to 9/5 Fahrenheit degrees. A temperature of x Celsius degrees is therefore $(9/5)x$ Fahrenheit degrees above the freezing point of water, i.e., $(9/5)x + 32$ degrees Fahrenheit. That is, the equation that yields the temperature y degrees Fahrenheit corresponding to x degrees Celsius is:

temperature in °F → $$y = \frac{9}{5}x + 32 \qquad ①$$ ↑ x: temperature in °C

To find an equation that converts Fahrenheit to Celsius degrees, we let x represent the temperature in °F and y the corresponding temperature in °C. The equation above yields:

temperature in °F → $$x = \frac{9}{5}y + 32$$ ↑ y: temperature in °C

Solving for y yields

temperature in °C → $$y = \frac{5}{9}(x - 32) \qquad ②$$ ↑ x: temperature in °F

Equations ① and ② are said to be inverses of each other. How are their slopes related?

WORKED EXAMPLE 1

a) Graph the linear function $y = \frac{9}{5}x + 32$ that gives the temperature y °F corresponding to a given temperature of x° Celsius for $0 \le x \le 100$.

b) Use your graph to determine the Fahrenheit temperature corresponding to 36°C.

c) Graph the linear function that gives the temperature y °C corresponding to a temperature of x°F. Use your graph to determine the temperature in Celsius degrees corresponding to a body temperature of 98.6°F.

d) Graph the linear function $y = \frac{9}{5}x + 32$ and its inverse for the range of values $-100 \le y \le 100$. State the intercepts and the slope of each graph.

Solution

a) To define the given linear function, we press:

Y= **9** **÷** **5** **X,T,θ,n** **+** **3** **2** **ENTER**

To graph it in the domain $0 \le x \le 100$, we press **WINDOW** and set Xmin and Xmax as shown. Then we press **ZOOM** **0** to graph the function in the appropriate window. This yields the graph shown in the display on the right.

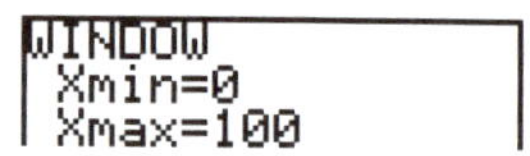

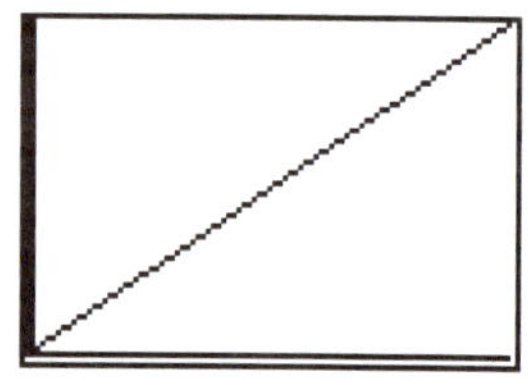

b) To determine the value of y corresponding to $x = 36$, we press:

2nd [CALC] **ENTER**

In response to the prompt for X, we enter the value 36 and we obtain the display shown here. The screen indicates that $y = 96.8$ when $x = 36$. That is, a temperature of 36°C corresponds to a temperature of 96.8°F. The temperature at Celsius' house on that day was 96.8°F!

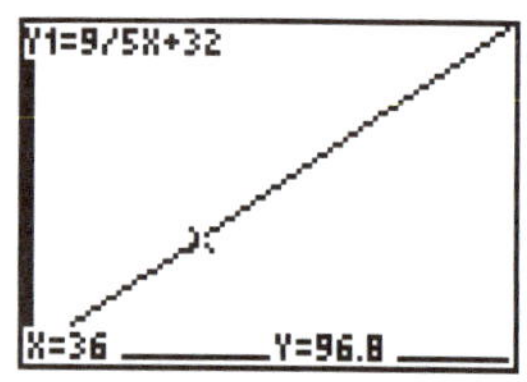

c) To graph the linear function that converts Fahrenheit to Celsius degrees, we can interchange the roles of x and y in the equation above and solve for y as on the previous page to obtain the linear function $y = \frac{5}{9}(x - 32)$. Then we graph this function Y_2 as in Part a) above with $32 \le x \le 212$ and proceed as in Part b) to obtain the display shown here, indicating that a body temperature of 98.6°F is 37°C.

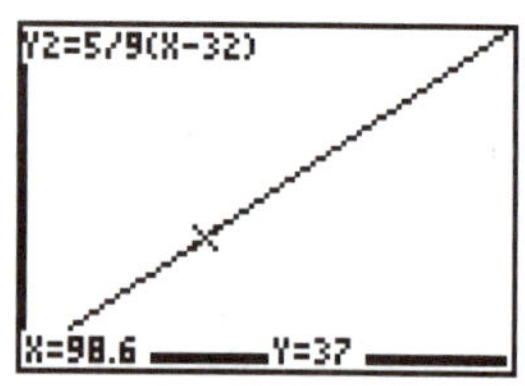

d) The graphs of the linear function and its inverse look the same because they are drawn in windows that match their range from the freezing to the boiling points of water. To display them on the same graph, we enter their equations on the "Y=" list with both "=" signs highlighted as shown in the display .

The window variables are set to $-100 \le x \le 100$; $-100 \le y \le 100$. We place tick marks 10° apart on the x-axis and y-axis by setting Xscl = 10 and Yscl = 10. To show the true slopes of these lines, we press **ZOOM** **5** and we obtain this display. We could trace along the graph to verify that the steeper line is $Y_1 = \frac{9}{5}x + 32$. Setting $x = 0$, shows its y-intercept is 32. Solving $\frac{9}{5}x + 32 = 0$ shows its x-intercept is $-160/9$. Therefore its slope is 32/(160/9) or 9/5 – the coefficient of x! Similarly, we can show that Y_2 has x- and y-intercepts of 32 and $-160/9$ respectively and a slope of 5/9.

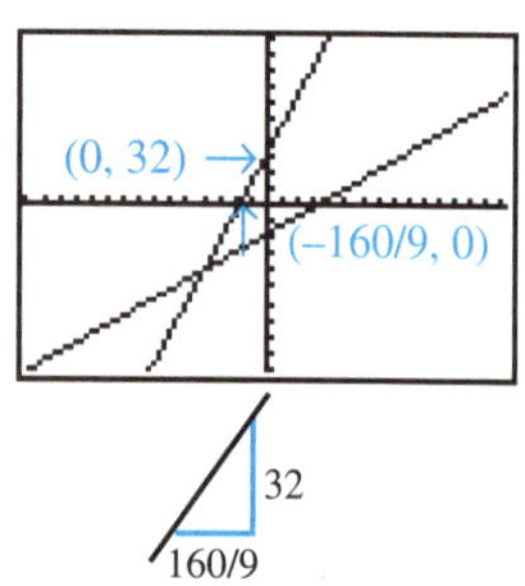

WORKED EXAMPLE 2

Graph the equation $3x + 7y = 8$, and state its slope and its x- and y-intercepts.

Solution

To graph this equation, we solve for y in terms of x. We write $y = \frac{8-3x}{7}$, i.e., $y = -\frac{3}{7}x + \frac{8}{7}$. Substitution of $x = 0$ into this equation yields the y-intercept $\frac{8}{7}$. To obtain the x intercept, we set $y = 0$ and solve for x to get $x = \frac{8}{3}$. The slope is given by $\frac{rise}{run} = \frac{-8/7}{8/3}$ or $-\frac{3}{7}$.

We could have deduced the slope and y-intercept directly once we expressed the equation in the form $y = -\frac{3}{7}x + \frac{8}{7}$. When the equation of a straight line is written in this form, (called the *slope-intercept form*) the coefficient of x gives the slope, and the constant term gives the y-intercept.

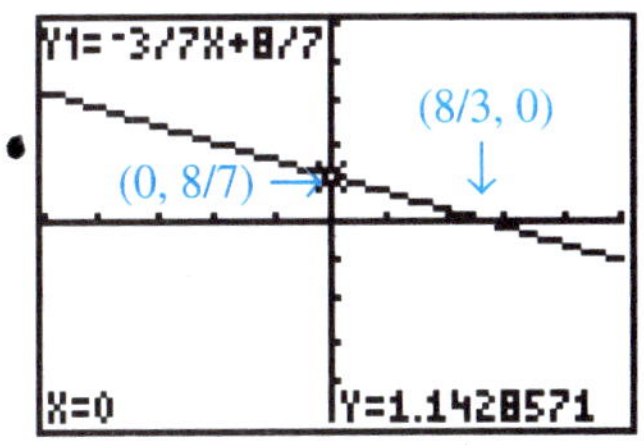

Worked Example 2 shows the following general result.

> The graph of the linear function defined by the equation $y = mx + b$, where m and b are constants and x is the variable, is a straight line with slope m and y-intercept b and conversely, a straight line with slope m and y-intercept b is the graph of the linear function defined by the equation $y = mx + b$.

WORKED EXAMPLE 3

a) Graph in the interval $-10 \leq x \leq 10$; $-10 \leq y \leq 10$, the straight line containing the points $(-7, 1)$ and $(3, -8)$.
b) Verify that the points $(-7, 1)$ and $(3, -8)$ lie on the graph.

Solution

a) To find the equation of the straight line passing through these two points, we calculate the slope of the line segment joining $(-7, 1)$ and $(3, -8)$.

$$slope = \frac{rise}{run} = \frac{1-(-8)}{-7-3} \text{ or } -\frac{9}{10}$$

Substitution of $m = -\frac{9}{10}$ into the general slope-intercept equation $y = mx + b$, yields the equation $y = -\frac{9}{10}x + b$. To determine the value of the constant b, we note that $(-7, 1)$ lies on the line and must therefore satisfy the equation, i.e., $1 = -\frac{9}{10}(-7) + b$. So, $b = -\frac{53}{10}$. The equation of the line is $y = -\frac{9}{10}x - \frac{53}{10}$. Graphing this line as in the worked examples above, yields the graph shown here.

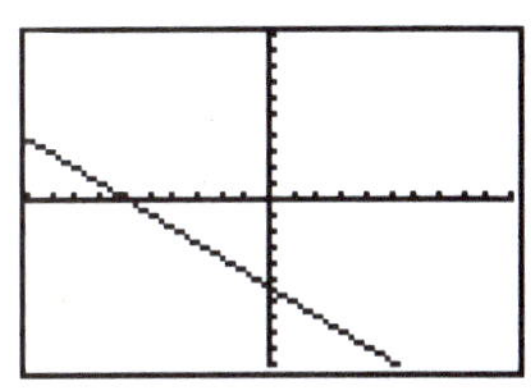

b) To determine the y-coordinate for $x = -7$, we press **2nd** [CALC] **ENTER** and enter -7 in response to the prompt. The display shows 1 indicating that $(-7, 1)$ is on the straight line. Similarly we verify that $(3, -8)$ is also on the line.

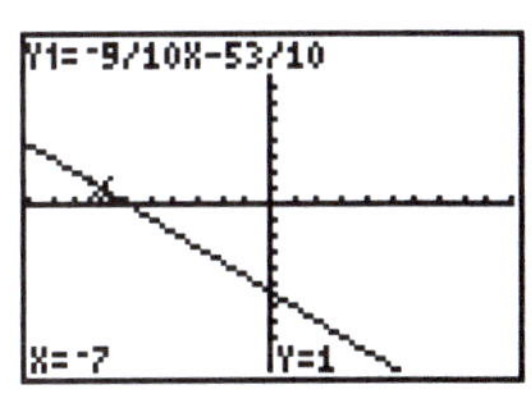

Exercises

1. Graph each of the following equations.

a) $y = 4x - 6$ b) $y = 1.4x + 8$

c) $2x - 5y = -6$ d) $3x + 5y - 2 = 0$

2. Write the slope and the y-intercept of each of the lines defined by the equations in *exercise* **1.**

3. Use the TRACE and the cursor keys to find the approximate *roots* of the equations in *exercise* **1**. (That is, find in each equation the approximate value of x corresponding to y = 0.)

4. Solve algebraically the equations in *exercise* 1 to check the approximate roots in *exercise* 2.

5. Any equation of the form $Ax + By + C = 0$ is called a *linear* equation because its graph is a straight line. Does every linear equation have a root? Why?

6. Make a table like this. Complete your table

Equation of the Line	Slope of the Line	Equation of Inverse	Slope of the Inverse Line
$y = 3x + 7$			
$y = 3x - 7$			
$3x + 2y = 6$			
$8x - 2y + 7 = 0$			

What is the relationship between the slope of the graph of a linear function and the slope of the graph of its inverse?

7. Write the equation of the line through each pair of points.

a) (-4, 6) and (7, 2) b) (-3,-8) and (3,7)

c) (-5, 5) and (6, -6) d) (-9,-9) and (1,1)

Graph each of the lines and verify that each of the given points is on the line.

8. a) Graph the lines in *worked example* 1 that define the relationship between degrees on the Fahrenheit and Celsius scales.

b) Convert 20°C and –10°C to Fahrenheit degrees.

c) Convert 75°F and 0°F to Celsius degrees.

d) What are the coordinates of a point that lies on both graphs? What is the physical significance of this point?

Investigations

9. Graph the equation $y = 2x$. Then press TRACE on your TI-84 Plus. You will observe a flashing cursor near the origin (0, 0). The coordinates of the cursor are shown on the screen.

Pressing the ▶ key several times in succession, moves the cursor to the right along the curve. Check that the coordinates of the points satisfy the equation $y = 2x$. What happens when you press the ◀ key several times?

Use the cursor keys to find the coordinates of any three points A, B and C on the graph defined by $y = 2x$. Record the coordinates of A, B and C and the slopes of AB and BC. What did you discover? What are the coordinates of the point at which the graph of the equation $y = 2x$ intersects the y-axis?

10. Repeat the investigation in **9** for the equation $y = 2x - 7$. In what way are the graphs of $y = 2x$ and $y = 2x - 7$ alike? different?

11. Repeat the investigation in **10** for the pair of equations $y = \frac{3}{2}x + 5$ and $y = \frac{3}{2}x - 3$ indicating how the graphs are alike and how they differ.

12. Generalize your discoveries in **10** and **11** to determine, (without graphing) the slope of the line with equation $y = \frac{5}{8}x - 4$ and the point at which the line intersects the y-axis.

CHALLENGE

L_1: $y = mx + b$ and L_2: $y = nx + c$ define two perpendicular lines.

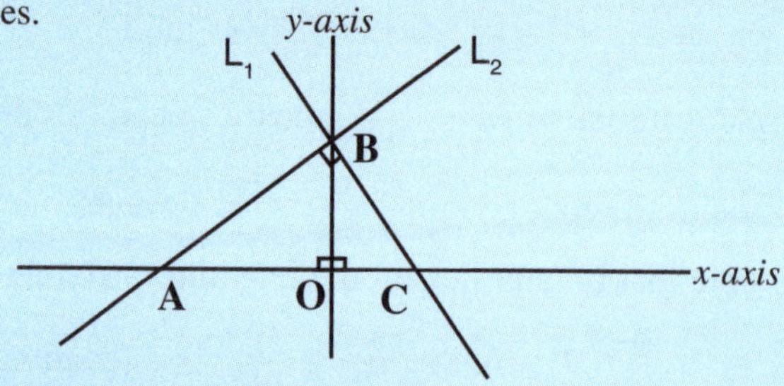

- Explain why ∠BAO = ∠CBO.
- Explain why the sides of ΔBAO are proportional to the sides of ΔCBO
- Calculate the slopes of L_1 and L_2 and state how the slopes of perpendicular lines are related.

Investigation: Linear Functions

WHERE SHOULD THEY HOLD THE FUNDRAISING PARTY?

Karen, Jason and their committee are planning a fundraising event to raise money for charity. They want to rent a hall and provide food and entertainment. Everyone in attendance will pay an admission fee to cover costs and the profit will be contributed to a local charity. To minimize the cost of the hall, Karen and Jason call three establishments to obtain quotes. Study each of the quotes, then complete the exercises.

1. Write an algebraic expression that expresses (as a function of x) the cost of providing a room and refreshments for x guests at each of these locations.

a) Leisure Lodge b) Cedar Glen Country Club c) Mariners' Inn

2. Graph each of the cost functions in exercise 1 for the range $0 \le x \le 90$ where x represents the number of guests.

3. List the coordinates of the points at which the graphs intersect and explain the significance of the points of intersections of these graphs.

4. Record for each location the range of values of x for which that location represents the least expensive alternative.

5. After Karen and Jason computed the costs, they contacted the three locations again to negotiate a better price. Each location agreed to reduce the per guest cost by \$2 but to keep the flat fees and minimum charges the same as before. That is, Leisure Lodge reduced its per guest cost to \$10 for those in excess of 25; Cedar Glen reduced its per guest cost to \$12 and the Mariners' Inn reduced its per guest cost to \$15.

Record for each location the new range of values of x for which that location represents the least expensive alternative. What is the least expensive location if 64 guests attend the event?

EXPLORATION 4 The Quadratic Function

Over 2500 years ago, the ancient Greeks explored the esthetics of geometric forms. Their search for ideal proportions led to the exploration of rectangular shapes and the investigation of problems such as the following.

What is the length-to-width ratio of the rectangle such that the removal of a square leaves a rectangle of the same shape as the original? (See diagram)

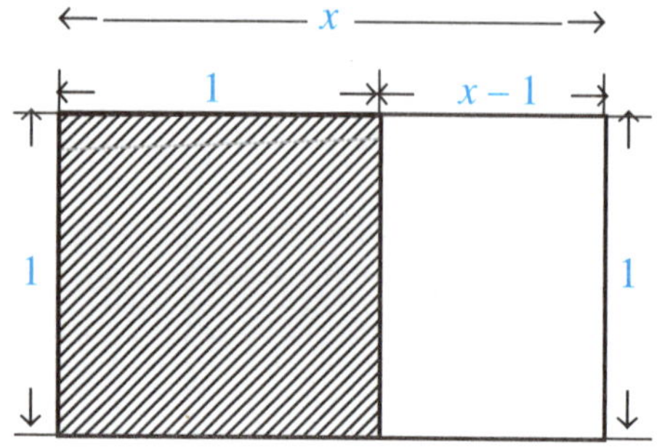

The answer to this problem is a number called the *golden ratio*– the most prominent ratio in mathematics next to π. It is also evident in the pyramid of Gizeh, the proportions of the Parthenon and the art of Leonardo da Vinci.

In modern algebraic notation, the problem would be expressed as:
What is the value of x such that $\frac{x}{1} = \frac{1}{x-1}$?

Although so-called *modern* algebraic notation would not be developed for another 2000 years, the ancient Greeks were able to solve this problem using their remarkable capacity for geometric deduction. (See *exercise 4*)

The Parthenon of ancient Greece was completed c. 438 B.C. Its front elevation displays the shape of a golden rectangle.

Cross-multiplication of the equation above yields the equation $x^2 - x - 1 = 0$. In high schools, the formula for the roots of quadratic equations is derived by completing the square. A more geometrically intuitive approach to the solution of this particular equation is given below and this method is generalized in *exercise 7.*

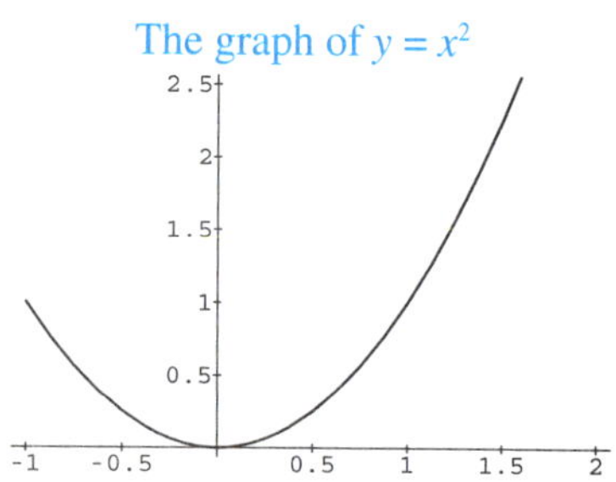

The graph of the equation $y = x^2$ is called a *parabola.* We see in the display that this parabola is symmetic about the y-axis (called its *axis of symmetry*) and it has a vertex at (0, 0). Furthermore, any point on the graph has the property that its height above the vertex is the square of its distance from its axis of symmetry.

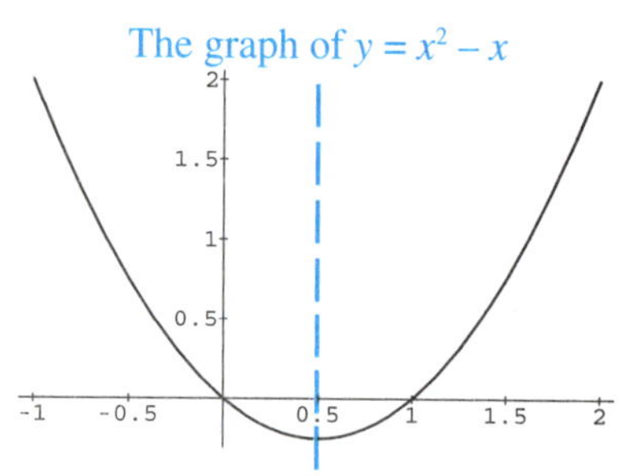

The graph of $y = x^2 - x$ is the same shape as the graph of $y = x^2$ except it is shifted to the right by 1/2 and downward by 1/4 unit. To see why this is so, we observe that $x^2 - x = x(x-1)$ so, this expression takes the value 0 for $x = 0$ and for $x = 1$. That is, the points (0, 0) and (1, 0) must lie on the graph of $y = x^2 - x$. The axis of symmetry must be $x = 1/2$ and the vertex must be (1/2, –1/4). Why?

Similarly, the graph of $y = x^2 - x - 1$ is the same shape as the graph of $y = x^2 - x$ except it is shifted downward by 1 unit. Therefore its axis of symmetry remains $x = 1/2$ and its vertex becomes (1/2, –5/4). Since the shape is preserved, the height of any point above the vertex is the square of its distance from the line $x = 1/2$. The points where the graph of $y = x^2 - x - 1$ intersects the x-axis (called *zeros of the function* $y = x^2 - x - 1$) are 5/4 units above the vertex, so if d denotes their distance from the line $x = 1/2$, then $d^2 = 5/4$, i.e. $d = \sqrt{5}/2$. Therefore the zeros are:

$$\left(\frac{1-\sqrt{5}}{2}, 0\right) \text{ and } \left(\frac{1+\sqrt{5}}{2}, 0\right)$$

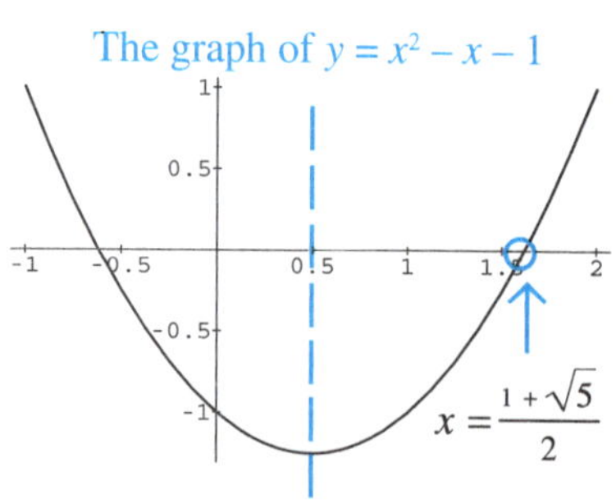

The roots of the equation $x^2 - x - 1 = 0$ are therefore $x = \frac{1-\sqrt{5}}{2}$ and $x = \frac{1+\sqrt{5}}{2}$.

The first of these is negative, and the second is the golden ratio, often denoted by τ.

WORKED EXAMPLE 1

Graph the quadratic function defined by the equation $y = 3x^2 - 6x + 2$.

a) Write the coordinates of the point on the graph with x-coordinate 2.5.
b) Write the coordinates of the point(s) on the graph with y-coordinate 7.
c) Write the coordinates of the vertex and the equation of the axis of symmetry.
d) Write the coordinates of the zeros and roots of the corresponding equation $3x^2 - 6x + 2 = 0$.

Solution

a) To define $Y_1 = 3x^2 - 6x + 2$, we press:

[Y =] [3] [X,T,θ,n] [x^2] [–] [6] [X,T,θ,n] [+] [2] [ENTER]

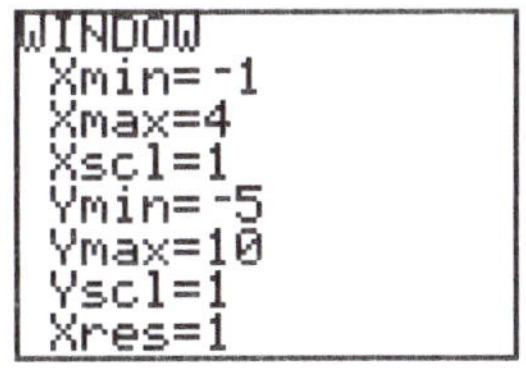

Pressing [ZOOM] [6] graphs Y_1 in the window $-10 \le x \le 10$; $-10 \le y \le 10$. To center the graph in the area of interest, we enter [WINDOW] and set the window variables as shown in the top display. Pressing [GRAPH] yields the graph shown. To find the coordinates of the point with x-coordinate 2.5, we enter:

[2nd] [CALC] [ENTER] 2.5 [ENTER]

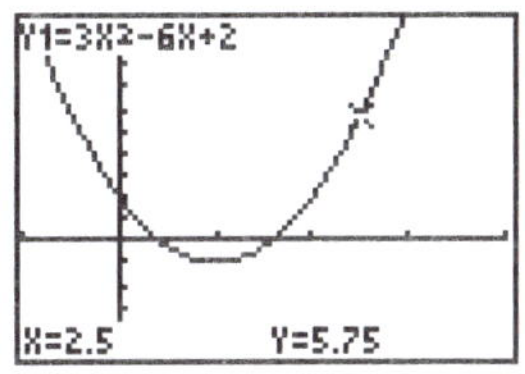

This produces the display indicating that the point (2.5, 5.75) is on the graph.

b) To determine the point with y-coordinate 7, we define $Y_2 =$ 7 and press [GRAPH]. To determine the point of intersection of the graphs defined by Y_1 and Y_2, we press

[2nd] [CALC] [5]

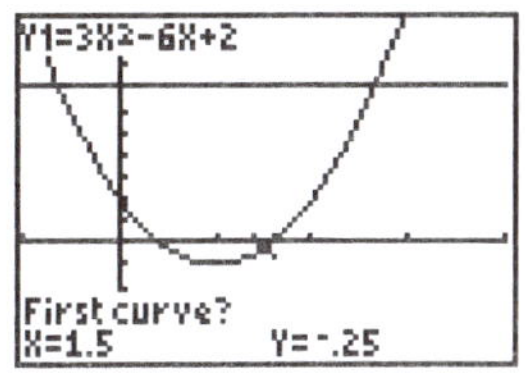

and we obtain the display shown here prompting us to identify the first of the two curves for which we seek the intersection. Pressing [ENTER] produces a prompt for us to identify the second curve. Pressing [ENTER] a second time yields a request (**Guess?**) for us to move the cursor close to the point nearer the point of intersection we wish to estimate. Pressing [ENTER] once more yields the display opposite. This tells us that (2.632…, 7) is one point on the graph. In the same way, we find the other intersection point is at (–0.632…, 7).

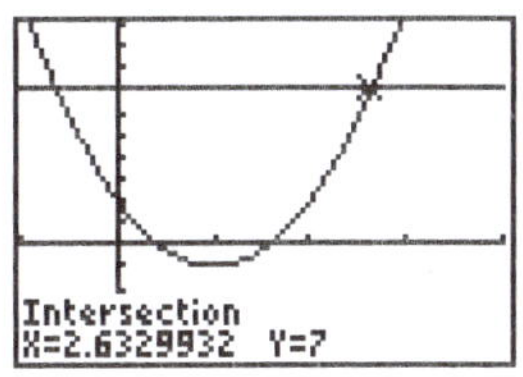

c) To obtain the coordinates of the vertex we enter: [2nd] [CALC] [3]
In response to the prompt **Left Bound?**, we trace to a point left of the vertex and press [ENTER]. In response to the prompt **Right Bound?**, we trace to a point right of the vertex and press [ENTER]. We obtain the display showing that a the vertex is the point (1, –1), so the axis of symmetry has equation $x = 1$.

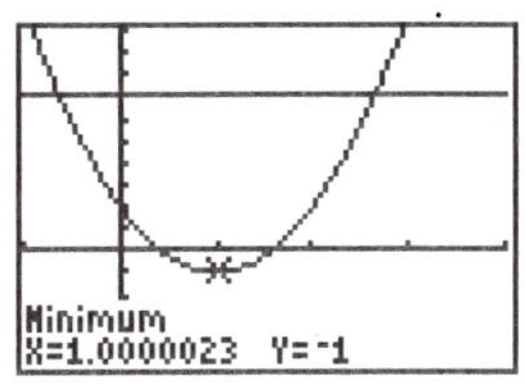

d) To determine the zeros of Y_1, we press [2nd] [CALC] [2]. In response to the prompts **Left Bound?**, and **Right Bound?**, we can trace as in part c) or merely enter 0 and 1 in turn. Pressing [ENTER] yields the display showing that one zero is at (0.422…, 0). Similarly we proceed to obtain the other zero (1.57…,0). Therefore the roots are $x = 0.422\ldots$ and $x = 1.57\ldots$.

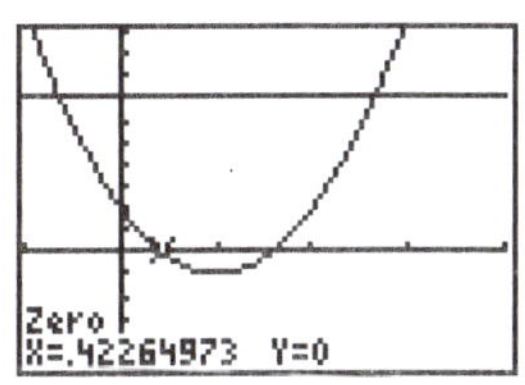

WORKED EXAMPLE 2

The profit y (in dollars) per CD sold is given by the equation $y = -97x^2 + 3800x - 23000$ where x is the retail price per CD

a) What is the profit when the retail price is \$16.95?

b) What retail price will yield a profit of \$10,000?

c) What retail price generates the largest profit?

d) What range of retail prices will yield a profit for the CD producer?

Solution

a) To enter the expression for the profit as a function of the retail price, we press:

[Y =] [(-)] 97 [X,T,θ,n] [x^2] [+] 3800 [X,T,θ,n] [–] 23000 [ENTER]

We then press [WINDOW] and set Xmin = 0; Xmax = 40 (the cost range for CDs) and to set the range for y we enter: [ZOOM] [0]

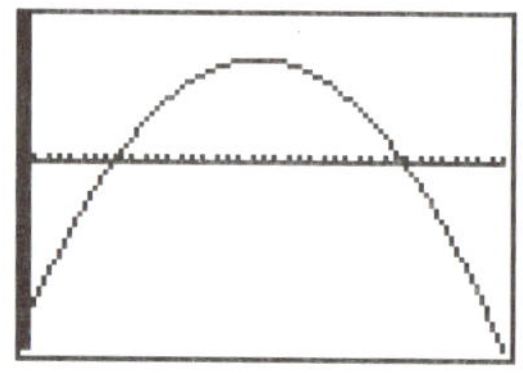

This sets Ymin = –26200 and Ymax = 14216.47…
To center the graph we set Ymin = –20000 and Ymax = 20000 and press [GRAPH]
We obtain the graph shown in the display. To determine the profit corresponding to x = \$16.95, we press [2nd] [CALC] [ENTER] 16.95 [ENTER] and we obtain this display. It shows that when the retail price is \$16.95, the profit is about \$13,500.

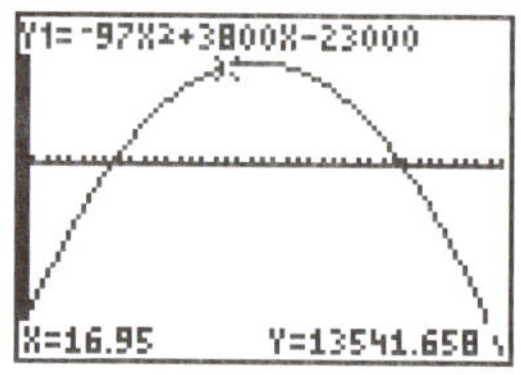

b) To determine the retail price that will yield a profit of \$10,000, we define Y_2 = 10000 and press [GRAPH]. To determine the point of intersection of the graphs defined by Y_1 and Y_2, we press [2nd] [CALC] [5] and we obtain the display shown here prompting us to identify the first of the two curves for which we seek the intersection. Pressing [ENTER] produces a prompt for us to identify the second curve. Pressing [ENTER] again yields a request (**Guess?**) for us to move the cursor close to the point nearer the point of intersection we wish to estimate. Pressing [ENTER] once more yields the display opposite. This tells us that a retail price of \$26.18 yields a profit of about \$10,000. In the same way, we find the other intersection point at a retail price of \$12.99.

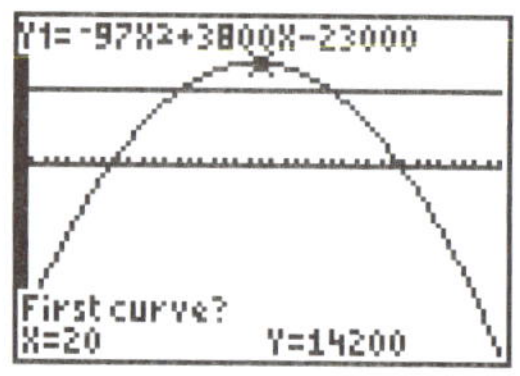

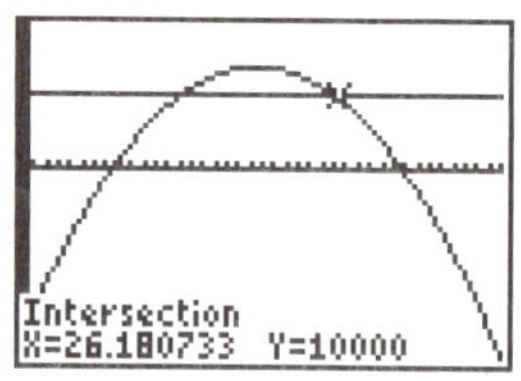

c) To find the retail price yielding maximum profit, we press:

[2nd] [CALC] [4]

In response to the prompt **Left Bound?**, we trace to a point left of the vertex and press [ENTER]. In response to the prompt **Right Bound?**, we trace to a point right of the vertex and press [ENTER]. We obtain the display showing that a retail price of \$19.59 yields a maximum profit of about \$14,216.

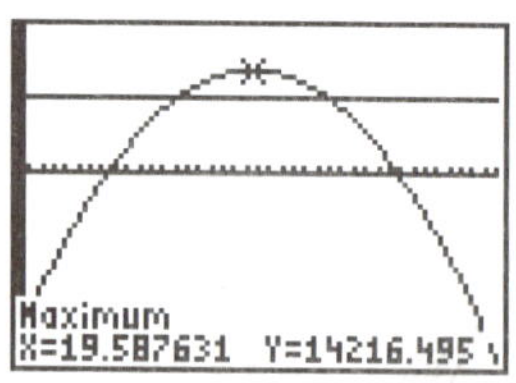

d) The retail prices yielding a profit is the set of values of x for which $y > 0$. This is the set of values of x between the zeros of Y_1. To determine the zeros of Y_1, we press [2nd] [CALC] [2]. Proceeding as in Part c) we find that the range of prices that yield a profit is $\$7.49 \le x \le \$31.69$.

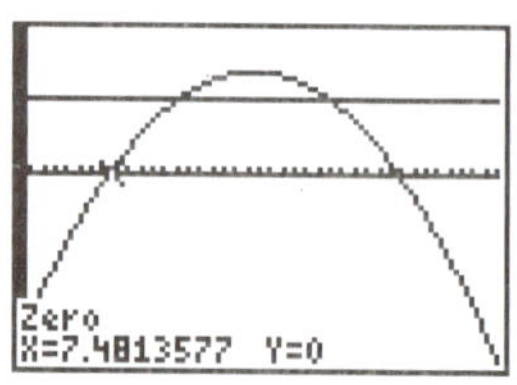

Exercises

1. For each of the functions below, write:

- the coordinates of the vertex,
- the equation of the axis of symmetry,
- the zeros
- the range (i.e., the set of values of y)

a) $y = x^2 - 4x - 21$ b) $y = 6x^2 + 33x + 42$
c) $y = x^2 + 2x - 5$ d) $y = -5x^2 - 2x + 7$

2. In *exercise 6*, you will show that the roots of the general quadratic equation $ax^2 + bx + c = 0$ are given by:

$$x = \frac{-b \pm \sqrt{b^2 - 4ac}}{2a}$$

Use this formula to solve the following equations.

a) $x^2 - 4x - 21 = 0$ b) $6x^2 + 33x + 42 = 0$
c) $x^2 + 2x - 5 = 0$ d) $-5x^2 - 2x + 7 = 0$

Compare these answers with the zeros you found in *exercise* 1. Describe what you discover.

3. Graph each of the following functions. Use the keying sequence 2nd [CALC] 2 to determine the zeros of each function.

a) $y = x^2 - 4x - 21$ b) $y = -4x^2 + 12x - 9$
c) $y = x^2 + 7x + 5$ d) $y = x^2 + 6x + 14$

Explain any error messages that occur when you attempt to find zeros. Explain how the number of roots of the equation $x^2 - 4x - 21 = 0$ is related to the number of zeros of the function $y = x^2 - 4x - 21$. Do this for the other functions in this exercise.

4. The diagram shows how the ancient Greeks constructed a golden rectangle, starting with a 2 × 2 square, bisecting its base and describing an arc. Prove that the result is a golden rectangle.

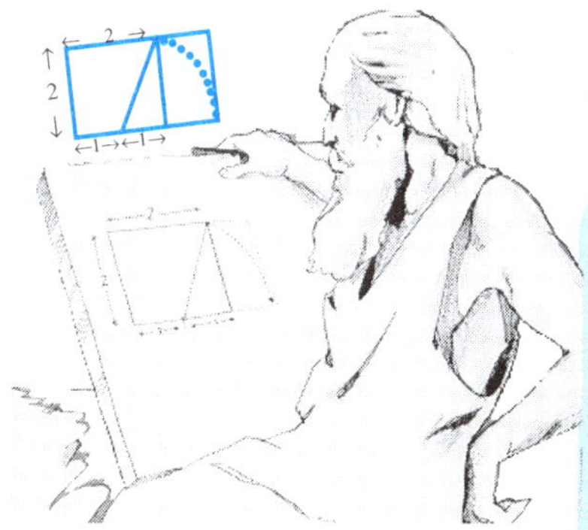

5. Given $\sigma = \frac{1-\sqrt{5}}{2}$ and $\tau = \frac{1+\sqrt{5}}{2}$;

a) prove $\sigma = -1/\tau$.
b) calculate $\tau^2, \tau^3, \tau^4, \ldots$
c) describe any pattern that you discover in part b) and use it to write an expression for τ^6.

Investigations

6. a) Prove that if $ax^2 + bx + c = 0$, then

$$x^2 + \frac{b}{a}x + \frac{b^2}{4a^2} + \frac{c}{a} = \frac{b^2}{4a^2}$$

b) Prove that if the equation in part a) is true then,

$$\left(x + \frac{b}{2a}\right)^2 = \frac{b^2 - 4ac}{4a^2}.$$

c) Prove that if the equation in part b) is true then,

$$x = \frac{-b \pm \sqrt{b^2 - 4ac}}{2a}.$$

d) Use parts a), b), and c) to prove that the roots of the equation $ax^2 + bx + c = 0$ are given by

$$x = \frac{-b \pm \sqrt{b^2 - 4ac}}{2a}.$$

7. In this investigation, you will derive a formula for the roots of the general quadratic equation $ax^2 + bx + c = 0$. For the quadratic function defined by each equation:

a) $y = ax^2$
b) $y = ax^2 + bx$
c) $y = ax^2 + bx + c$

- write the equation of the axis of symmetry of the graph of the function.
- write the coordinates of the vertex of the graph.
- for any point (x, y) on the graph, write its height above the vertex in terms of its distance d from the axis of symmetry.
- write the zeros of the function.

Using your answers to the questions above, write a formula for the roots of the equation $ax^2 + bx + c = 0$ in terms of a, b, and c.

CHALLENGE

When you press the Y = key, you see a display of ten or fewer variables which you have previously defined. These variables are denoted Y_1, Y_2, Y_3...and so on.

You can define a new variable in terms of these by using the VARS ▶ ENTER key combination. Define the variables $Y_1 = x^2$ and $Y_2 = 2x - 4$.

Then use the VARS ▶ ENTER sequence to define $Y_3 = Y_1 + Y_2$. Graph Y_1, Y_2 and Y_3. Write the equations of the axes of symmetry of Y_1 and Y_2. Redefine Y_2 by changing the coefficient of x. How is the axis of symmetry of the graph of $y = ax^2 + bx + c$ related to the coefficients a and b?

EXPLORATION 5 The Quest for Roots of Higher Order Equations

The study of algebra began as the quest for roots of equations. Cuneiform clay documents of the ancient Babylonians (c. 1700 B.C.) reveal that they were able to solve linear and quadratic equations expressed in rhetorical style. When algebraic symbolism was developed during the Italian Renaissance in the 16th and 17th centuries, those procedures for finding roots of linear and quadratic equations emerged as the familiar formulas that we studied in the previous explorations.

As early as 1100 A.D. Omar Khayyam, the Persian poet and mathematician, discovered a geometric procedure for finding the real roots of cubic equations. However, it was not until the 16th century that an algebraic formula was found for roots of the general cubic equation $ax^3 + bx^2 + cx + d = 0$. The origin of this discovery remains one of the great enigmas in the history of mathematics. The following sequence offers a brief overview of the events leading to the ultimate conquest in the search for roots of algebraic equations.

1

•1515 Scipioni del Ferro discovers how to solve equations of the form $x^3 + cx = d$. He reveals this discovery in secrecy to his student, Antonio Maria Fior.

•1535 Niccolo Tartaglia claims to have discovered how to solve equations of the form $x^3 + bx^2 = d$, so Fior challenges him to a public equation-solving duel.

•Tartaglia wins the duel having discovered only days earlier how to solve equations of the type known to Fior.

•Tartaglia confides his method to Professor Girolamo Cardano.

Nicolo Tartaglia 1499-1557

2

•1545 Girolamo Cardano publishes *Ars magna* in which he presents the formula for the roots of the general cubic equation, violating his pledge of secrecy to Tartaglia.

•Tartaglia protests vehemently, accusing Cardano of plagiarism. Ludovico Ferrari, a brilliant student of Cardano, defends his teacher, claiming that Cardano's source was Fior!

•*Ars magna* included also the solution to the general quartic equation discovered by that same Ferrari in 1540 when he was still a teenager!

Girolamo Cardano 1501-1576

3

•1750 Leonhard Euler fails to solve the general quintic equation, $ax^5 + bx^4 + cx^3 + dx^2 + ex + f = 0$.

•1824 Neils Abel proves that the roots of the general quintic equation cannot be expressed by means of radicals involving only the coefficients of the equation.

•1832 Evariste Galois dies at the age of 21 in a pistol duel over a young lady. His legacy of 63 pages, many of them frantically scrawled the night before the duel, contained for an algebraic equation of any degree necessary and sufficient conditions that it be solvable by radicals.

Evariste Galois 1811-1832

WORKED EXAMPLE 1

An ancient Babylonian tablet displays a table of the values of $n^3 + n^2$ for $n = 1$ through 30. Create such a table to find a root of the equation:

$$x^3 + 2x^2 - 3136 = 0$$

Solution

To create the required table, we press:

Y= X,T,θ,n ∧ 3 + 2 X,T,θ,n
x² − 3 1 3 6 2nd
[TBLSET] 1 ENTER 1 2nd [TABLE]

The display shows the table below left.

X	Y1
1	-3133
2	-3120
3	-3091
4	-3040
5	-2961
6	-2848
7	-2695

X=1

X	Y1
8	-2496
9	-2245
10	-1936
11	-1563
12	-1120
13	-601
14	0

X=14

Using the ▼ key, we scroll down to X = 14 where we observe $Y_1 = 0$, so 14 is a root of the equation.

WORKED EXAMPLE 2

Graph the curve defined by the equation:

$$y = 2x^3 - 6x^2 - x + 6$$

Then trace along the curve to find the approximate values of the roots of the corresponding equation.

$$2x^3 - 6x^2 - x + 6 = 0$$

Solution

To graph this equation, we press:

Y= 2 X,T,θ,n ^ 3 − 6 X,T,θ,n x^2 − X,T,θ,n + 6 ENTER GRAPH

The screen displays the graph in the display.

We observe that the graph crosses the x-axis three times, indicating that there are three real values of x for which y is zero.

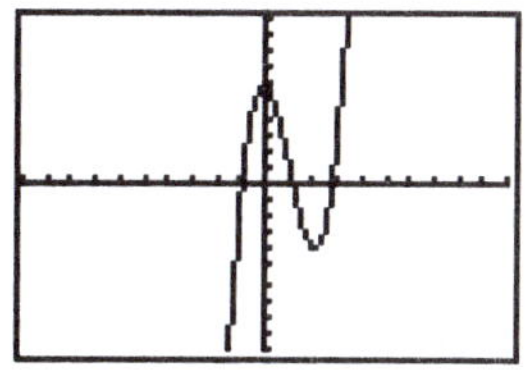

To approximate these roots, we press TRACE and observe a flashing cursor at the top of the "hill" or maximum of the curve. The screen indicates that the cursor is at the point on the curve with coordinates (0, 6).

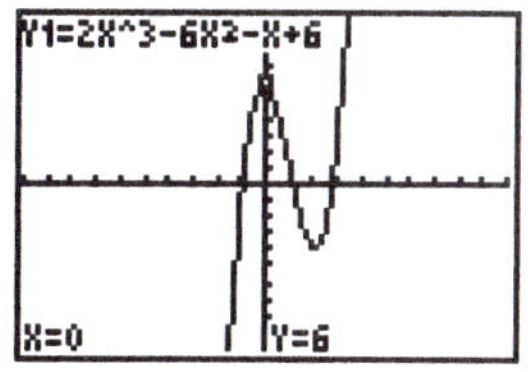

By pressing the cursor keys ◄ and ►, we can trace along the curve to the points where the curve crosses the x-axis. If we trace along the curve to the right, we reach the point (1.063…, .553…). The next point in the tracing is (1.276…, -.893…). Since the y-coordinate changes sign, we conclude that the curve crosses the x-axis somewhere between these two points. That is, if x_1 denotes the root, then $1.063 \le x_1 \le 1.276\ldots$

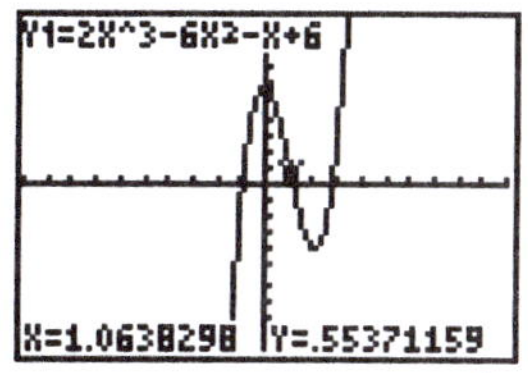

To obtain a closer approximation to this root, trace along the curve to the point (1.063…, .553…) and then press:

ZOOM 2 ENTER

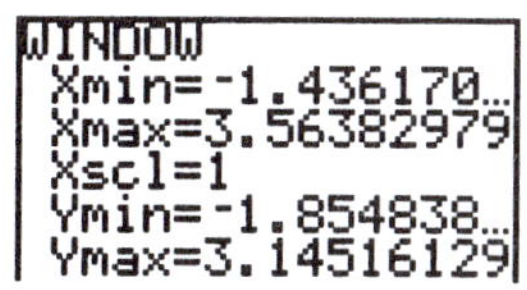

This keying sequence is called *zooming in* and it causes the curve to be redrawn with the window shrunk by a factor of 4;
i.e. $-1.43\ldots \le x \le 3.56\ldots$, $-1.85\ldots \le y \le 3.14\ldots$ (See the display).

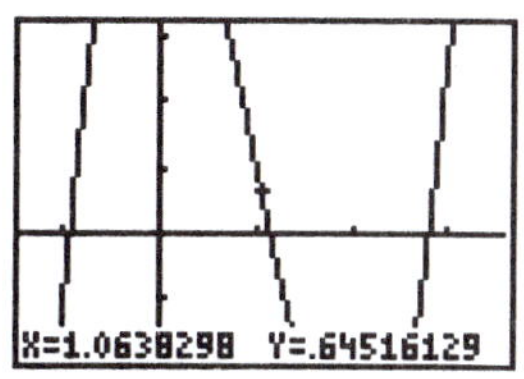

Proceeding as above, we obtain $1.1170.. \le x_1 \le 1.1702\ldots$
Repeated zooming yields closer approximations to x_1.
For approximations to the roots x_2 and x_3 see *exercise* 5.

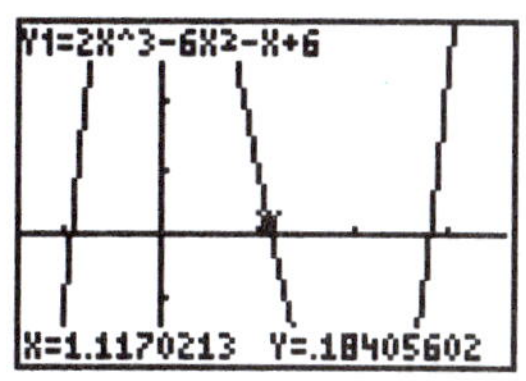

WORKED EXAMPLE 3

Determine the roots of the equation

$$2x^3 - 6x^2 - x + 6 = 0$$

correct to 6 decimal places.

Solution

We graph this equation using the keying sequence in *worked example* 2. It is clear from the graph that this equation has three real roots.

To determine the value of the smallest positive root, x_1, without using the repeated iterations in *worked example* 2, we press these keys:

2nd [CALC] **2**

Note: A left bound is any value of x which is less than x_1, so we could trace along the curve to any point left of $(x_1, 0)$ and press **ENTER**.

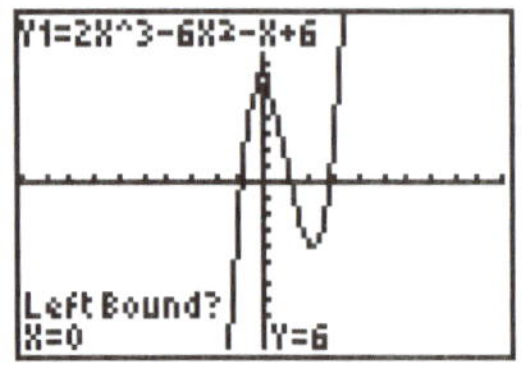

We obtain the display shown.

The prompt, **Left Bound?** requests that we enter a value of x which is left of the root, x_1,(i.e. $x < x_1$). Since $0 < x_1$, we press:

0 **ENTER**

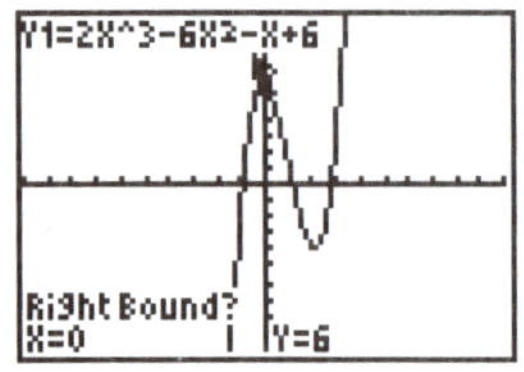

The display shows the prompt, **Right Bound?**. The scale markings on the x-axis show that $x_1 < 2$, i.e. 2 is on the right of x_1, so we press:

2 **ENTER**

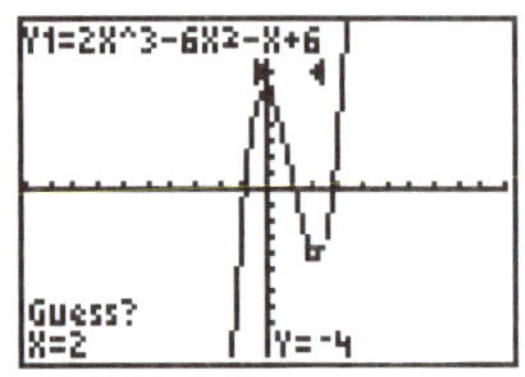

The display shows the prompt, **Guess?**. To enter this upper bound as our guess, we press the **ENTER** key one more time. The screen shows the cursor on the graph near the root and displays:

Zero
X = 1.143705   Y = 0

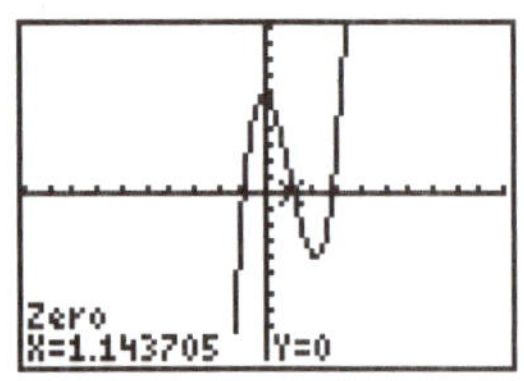

This indicates that when $x = 1.143705$, the value of the polynomial is close to zero. To verify that this value of x is an approximation to the actual root x_1 correct to 6 decimal places, we must verify that $1.1437045 \le x_1 < 1.1437055$. That is, we must show that the cubic polynomial changes sign (has a zero) in this interval.

To evaluate the polynomial at $x = 1.1437045$, we press:

2nd [CALC] **1**

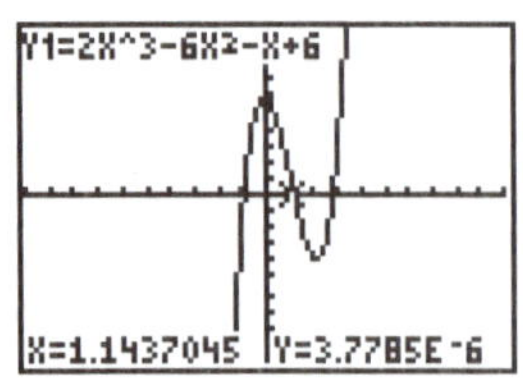

In response to the prompt, **X =** , we enter 1.1437045. The display shows the corresponding value of y is 3.7785 E -6.

Repeating this procedure, we find the value of y corresponding to 1.1437055 is -3.098 E -6. Since y changes sign in this interval, x_1 lies in this interval and $x_1 = 1.143705$ (to 6 decimal places).

Similarly, we determine x_2 and x_3. (See *exercise* 5.)

Exercises

1. Graph each of the equations below. Then use the TRACE and the cursor keys to find approximations to *all* the (real) roots of each equation.

a) $y = 3x^2 - 5$ b) $y = 3x^2 + 2$

c) $y = 2x^2 - 5x - 3$ d) $y = 2x^2 - 5x + 5$

Use the [CALC] menu to check your answers. What did you discover in parts "b" and "d". Explain what happened.

2. Compare the number of real roots of equation $3x^2 - 5 = 0$ with the number of real roots of the equation $3x^2 + 2 = 0$. Compare the number of real roots of equations $2x^2 - 5x - 3 = 0$ and $2x^2 - 5x + 5 = 0$.

Explain how adding a constant to one side of an equation changes the graph and possibly the number of real roots of the resulting equation.

3. Graph each of the following functions in the window: $-5 \le x \le 5; -50 \le y \le 50$.

a) $y = 6x^4 - 5x^3 - 18x^2 + 10x + 12$

b) $y = -6x^5 + 5x^4 + 18x^3 - 10x^2 - 12x$

c) $y = 21x^3 + 19x^2 - 61x + 21$

d) $y = -6x^4 + 11x^3 + 13x^2 - 16x - 12$

4. Use the TRACE and the cursor keys to find the approximate roots of the corresponding equations in exercise **3**. (That is, find for each equation the approximate value of x corresponding to $y = 0$.)

5. Approximate *all* real roots of each of the following equations to six decimal places.

a) $x^3 + x^2 - 3x - 1 = 0$

b) $2x^3 - 6x^2 - x + 6 = 0$

c) $\frac{1}{8}x^3 - x^2 + \frac{3}{2}x + 1 = 0$

d) $\frac{1}{2}x^4 - \frac{1}{2}x^3 - 5x^2 + 4x + 8 = 0$

Verify your answers as in *worked example* 3.

6. Write the equation of a function that has each set of zeros. Then graph that function in the window:
$-5 \le x \le 5; -50 \le y \le 50$.

a) $\{(-2, 0), (2, 0), (-1, 0)\}$
b) $\{(1, 0), (3, 0), (-\sqrt{2}, 0), (\sqrt{2}, 0)\}$
c) $\{(1 + \sqrt{3}, 0), (1 - \sqrt{3}, 0), (3, 0), (0, 0)\}$

Investigations

7. From your discoveries in *exercise* 5, conjecture answers to the following. (An equation of *degree n* is an equation in which the highest power of x is x^n.)

a) What is the maximum number of real roots which an equation of degree n can have?
b) What is the minimum number of real roots which an equation of degree n can have?
c) Is there a cubic equation that has no real roots? If so, give an example.

8. a) Graph the equation $y = \frac{1}{x^2} - 0.001$.

Use the TRACE key to find a value of x for which y is approximately 0. Trace along the curve to the right to find a value of x for which $y < 0.001$. Denote this value by x_0. Is x_0 close to the actual root of this equation?

b) Use the [CALC] menu as in *worked example* 3 to find a positive root of the given equation. Is x_0 close in value to the positive root?

c) If y is very close to 0 for a given value of x, is that value of x a close approximation to a root of the equation? Explain your answer using an example.

d) Solve the given equation algebraically. Compare the root found algebraically with the root found using the [CALC] menu. Why do we not solve all equations algebraically?

CHALLENGE

An ancient manuscript asks the edge length of a cube such that half its volume plus one sixth of its surface area is 4 units more than half the total length of all its edges. (Assume the edge length is x units and that the surface area and volume are measured in squared and cubed units respectively.)

a) Write an algebraic expression for y in terms of x if y is the difference between "half the volume plus one-sixth the surface area" and "half the total length of all its edges plus 4".

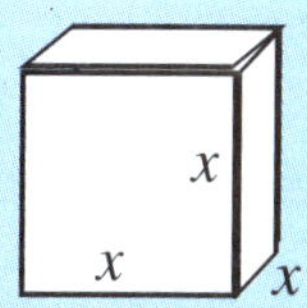

b) Graph the equation which expresses y in terms of x.

c) Approximate the real root to 2 decimal places.

EXPLORATION 6 Roots of Systems of Equations

How easy is it to gain entry to heaven? The cartoon suggests that it's not so easy if you're math phobic. However, recent reports indicate that graphing calculators have now been approved for use on all entrance examinations including those which govern the afterlife. This means that even the most phobic among us can solve problems of the type administered in the cartoon, by merely expressing the given relationships as equations and solving them. This procedure is modeled below.

THE FAR SIDE® By GARY LARSON

Math phobic's nightmare

WORKED EXAMPLE 1

A train leaves Philadelphia at 1:00 p.m. Another train leaves Denver at 4:00 p.m. destined for Philadelphia. If the distance by train between Philadelphia and Denver is 2000 miles, and the trains from Philadelphia and Denver average 65 m.p.h. and 85 m.p.h. respectively, at what time will they pass each other?

Solution

Let x denote the time in hours between the departure of the first train, A, from Philadelphia and its encounter with train B from Denver. Furthermore, let y denote the distance in miles between Philadelphia and the point at which the trains pass.

Train	Distance	Time	Rate
A	y	x	$\frac{y}{x}$
B	$2000 - y$	$x - 5$	$\frac{2000-y}{x-5}$

Philadelphia and Denver are in different time zones. At 4:00 p. m. Denver time, it is 6:00 p.m. Philadelphia time.

Using these variables and the information given in the problem, we complete the table on the right showing the time, rate and distance for each train from the time of departure until the moment of encounter.

The conditions that the average speeds are 65 m.p.h. and 85 m.p.h. respectively, can be written as the following equations:

$\frac{y}{x} = 65$ and $\frac{2000-y}{x-5} = 85$, which expressed in terms of y become

$$y = 65x$$
and
$$y = 2425 - 85x$$

This pair of linear equations in x and y is called a *system of equations.*

To *solve* such a system of equations means to find values of x and y which satisfy both equations. We can solve this system algebraically by substitution or graphically as follows:

Press **WINDOW** and set the range values to $0 \le x \le 20$ and $0 \le y \le 2000$. Then graph $Y_1 = 65x$ and $Y_2 = 2425 - 85x$ to obtain the display shown.

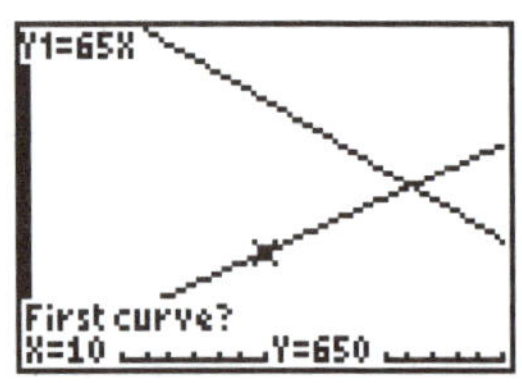

To obtain the point of intersection,press: **2nd** [CALC] **5** . In response to the prompts for **First Curve?, Second Curve?** and **Guess?**, press **ENTER** . The display shows **Intersection X = 16.16... Y = 1050.83...** That is, the trains pass at 5:10 a.m. Philadelphia time about 1050.83 miles from Philadelphia.

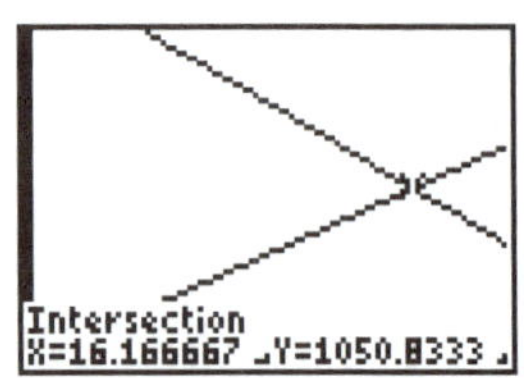

WORKED EXAMPLE 2

The two outermost planets in our solar system, Neptune and Pluto, travel in elliptical orbits around the sun which is located at the focus of both ellipses.

The orbit of Neptune is given by the Cartesian equations: $y = \pm\sqrt{904 + 0.54x - x^2}$

The orbit of Pluto is given by the Cartesian equations: $y = \pm 0.968\sqrt{1465 + 19.68x - x^2}$
where x and y are expressed in *astronomical units* (A.U.)

1 A.U. = distance from the earth to the sun

a) Graph the orbits of the planets and determine whether they intersect.
b) If the orbits intersect, determine the coordinates of the point of intersection.

Solution

a) The expression, $y = \pm\sqrt{904 + 0.54x - x^2}$, defines two functions, Y_1 and Y_2 where Y_1 and Y_2 are defined by:

$$Y_1 = \sqrt{904 + 0.54x - x^2} \text{ and } Y_2 = -\sqrt{904 + 0.54x - x^2}$$

To define Y_1, we press these keys:

[Y=] [2nd] [√] [9] [0] [4] [+]
[.] [5] [4] [X,T,θ,n] [−] [X,T,θ,n] [x^2] [)] [ENTER]

To define Y_2 as $-Y_1$, we press these keys:

[Y=] [(−)] [VARS] [▶] [ENTER] [1] [ENTER]

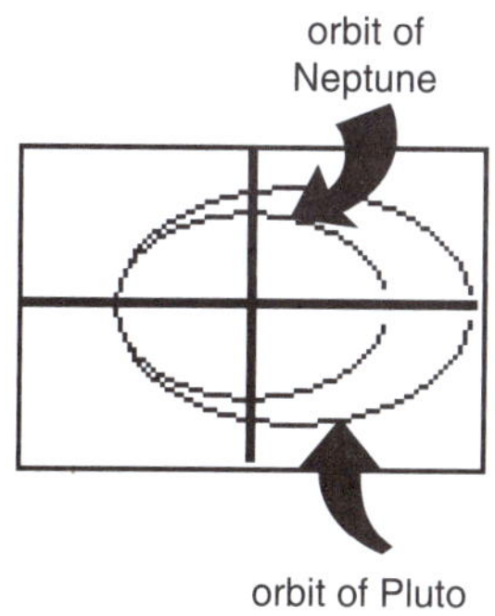

Similarly, define $Y_3 = 0.968\sqrt{1465 + 19.68x - x^2}$ and $Y_4 = -Y_3$.
The graphs of Y_1 and Y_2 together form the orbit of Neptune.
The graphs of Y_3 and Y_4 together form the orbit of Pluto.

To graph these orbits, press [WINDOW] and set the range values to $-50 \le x \le 50$ and $-50 \le y \le 50$. Then press [GRAPH] to obtain this display. Vertical distances on the TI-84 Plus display are slightly compressed relative to horizontal distances, so that circles will appear to be ellipses. To change the display to show the true shape of the orbits, press: [ZOOM] [5]. We obtain the display shown here.

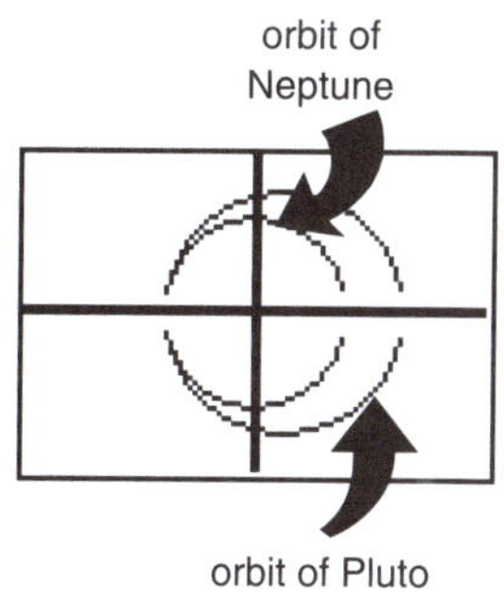

b) It is not clear from the display whether or not the curves intersect, so we solve for an intersection point by pressing:

[2nd] [CALC] [5]

The cursor appears on Y_1 and the prompt **First curve?** appears.
We press [ENTER] to input the first curve. The cursor moves to Y_2.

We press [▼] to move the cursor to Y_3, then press [ENTER] to input the second curve. We move the cursor near the intersection and press [ENTER] to obtain the intersection point in the display.

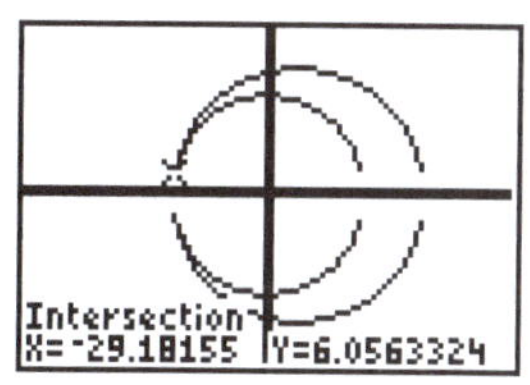

Exercises

1. Graph each pair of equations.

a) $y = 3x - 2$, $y = -2x + 7$

b) $15x - 12.3y = 21$, $17x + 16.2y = 47$

Solve using each of these methods:

- algebra
- tracing along either line to a point of intersection.
- using the intersect function on the CALC menu.

Which of these methods do you prefer? Explain.

2. Solve each system of equations algebraically.

a) $1.2x - 0.4y = 2.8$, $6x - 2y = 14$

b) $-6x + 14y = 17$, $3x - 7y = 13$

Graph each pair of equations. Use the intersect function on the CALC menu to find the point(s) of intersection. Explain what you discover.

3. Two poles of height 6 m and 24 m are 30 m apart. How high above the ground is the point of intersection of the lines running from the top of each pole to the foot of the other?

4. A family traveled a total distance of 740 km in 12 hours and 15 min on a trip from Toronto to Lewes, Delaware. Part of the trip was by car at 80 km/h and the remainder by ferry at 16 km/h. How many hours were spent traveling by car?

5. How many solutions has a pair of linear equations? Explain how you can determine the number of solutions by examining the graphs of the two equations.

6. Can you predict how many solutions to each system of equations without graphing?

a) $3x - y = 2$, $16x^2 + 49y^2 - 80x = 684$

b) $3x - y = -15$, $16x^2 + 49y^2 - 80x = 684$

Graph and solve each system using the CALC menu.

7. Solve this system of equations.

$$16x^2 + 49y^2 - 80x = 684$$

$$y = \frac{1}{8}x^3 - 2$$

What is the greatest possible number of points of intersection of a cubic polynomial and a quadratic polynomial? Give reasons for your answer.

Investigations

8. Graph these equations.

$$y = 3x - 4$$

$$y = 3x$$

$$y = 3x + 5$$

Explain how all three graphs are alike. How are they different? Describe what happens to the graph of the equation, $y = Ax + B$ when:

a) A is held constant and B is changed.

b) B is held constant and A is changed.

9. Graph each system and predict the number of solutions. Then solve using the CALC menu. (Set the window values to $-20 \le x \le 20$ and $-20 \le y \le 20$.

a) $83x - 48y = 400$, $9x^2 + 3y^2 - 7x = 382$

b) $83x - 48y = 799$, $9x^2 + 3y^2 - 7x = 382$

10. Graph each system and predict the number of solutions. Then solve using the CALC menu.

a) $3x - 5y = 6$, $9x^2 - 36x + 25y^2 = 108$

b) $3x - 5y = 6$, $9x^2 - 36x - 25y^2 = 108$

11. The equations of the orbits of the earth and Venus around the sun (at the origin) are given by:

earth: $x^2 - 0.034x + y^2 = 1$

Venus: $x^2 - 0.011x + y^2 = 0.52$

where all distances are given in astronomical units. Graph these orbits.

Press: ZOOM 2 to zoom in on the graphs.

Use ZOOM 5 to see the actual shapes of the orbits. Do the orbits of Venus and earth intersect?

Will Neptune and Pluto Collide?

Research the motions of the planets around the sun. We learned in worked example 2 that the orbits of these two planets appear to intersect. Write a report which indicates whether these planets will ever collide and give reasons for your answer.

EXPLORATION 7: Expanding the Concept of f(x)unctions

On page 7, we reviewed the different definitions of a function as the function concept evolved from an algebraic expression relating two variables to a set of ordered pairs with first components all distinct. We recall the first definition, attributed to Leonhard Euler.

> **•1734** *A function, f(x), is any algebraic expression involving variables and constants defined by an equation or graph.*

To generalize the function concept so that it would apply to not only algebraic but also non-algebraic (i.e. transcendental) expressions, Dirichlet introduced the following definition.

> **•1837** *If two variables, x and y are so related that whenever a value is assigned to x there is automatically assigned by some rule of correspondence, a value to y, then we say y is a "function" of x.*

The Dirichlet definition depends upon the concepts of "number" and "variable". Near the end of the 19th century, mathematicians sought a definition of function that could also have meaning outside the realm of numbers.

In 1874 German mathematician Georg Cantor published his revolutionary paper on *Mengenlehre* – a theory of sets. In this paper, he defined sets to have the same number of members if they could be placed in one-to-one correspondence. This definition presented no problem for finite sets, but it created paradoxical results for infinite sets. For example, it implied that the set of even integers was in some sense equal in size to the set of *all* integers. Cantor's paper spawned bitter disputes in the mathematics community. Cantor himself suffered a nervous breakdown admidst all the controversy. Fortunately, he lived long enough to see *Mengenlehre* celebrated as a significant contribution to mathematics. This work provided a framework for the modern definition of a function.

Georg Cantor 1845 – 1918

> *A function, f, is any set of ordered pairs of elements such that if $(x_1,y_1) \in f$, $(x_2, y_2) \in f$ and $x_1 = x_2$, then $y_1 = y_2$.*

WORKED EXAMPLE 1

Graph the function defined by $f(x) = x^2 + x + 12$.
State the domain and range of this function.

> **Note:**
> In the graph of a function, the top of a "hill" and the bottom of a "valley" are referred to respectively as a *local maximum* and a *local minimum*. The *extrema* of a function is the set of all its local maxima and minima.

Solution

Before graphing a function we press: **ZOOM** **6** to restore the standard window defined by $-10 \le x \le 10$; $-10 \le y \le 10$.

To graph the function $f(x)$, we press the following keys.

Y= **X,T,θ,n** **x²** **+** **X,T,θ,n** **+** **1** **2** **ENTER** **GRAPH**

Nothing appears on the screen. This indicates that the graph of $f(x)$ has no points that lie within the standard window.

To expand the window by a factor of 4, we press:
ZOOM **3** **ENTER** to obtain the graph shown.

The graph is a parabola with a minimum near $x = 0$.
To estimate the y-coordinate at this minimum, we press:

2nd [CALC] **3**

In response to the prompt **Left Bound?**, press: **(–)** **1** **ENTER**.

In response to the prompt **Right Bound?**, press: **1** **ENTER**.

In response to the prompt **Guess?**, press: **ENTER**.

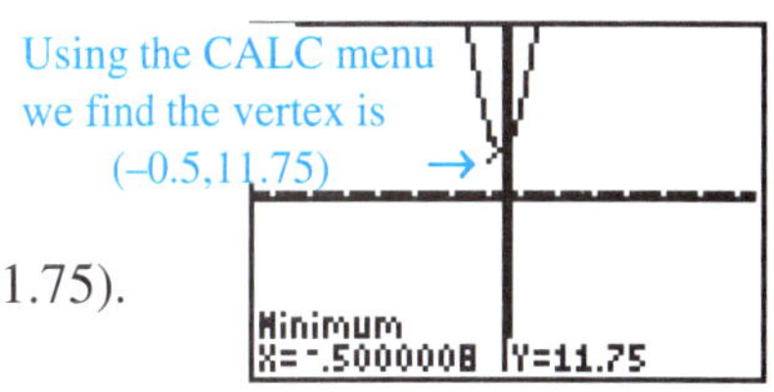

We find that the minimum has coordinates approximated by (-0.5000006, 11.75).
The domain of $f(x)$ is the set of all real x; the range is $\{ y \mid 11.75 \le y \}$.

WORKED EXAMPLE 2

Graph the function defined by $f(x) = 2x^3 - 7x^2 + 3x + 1$.
Determine the values of x at which $f(x)$ is zero and at which $f(x)$ has its extrema.

Solution

Press **Y =** then enter the expression for $f(x)$ and press **GRAPH**.
To find the points at which $f(x)$ is zero (i.e. roots of $f(x) = 0$):
•Press **2nd** [CALC] **2** to get the graph in the display.

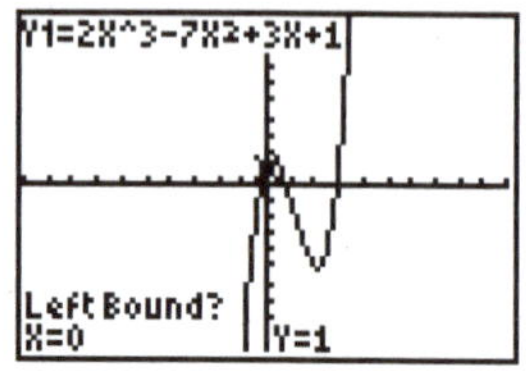

•In response to the prompt **Left Bound?**, press: **(–)** **1** **ENTER**.

•In response to the prompt **Right Bound?**, press: **0** **ENTER**.

•In response to the prompt **Guess?**, press: **ENTER**.

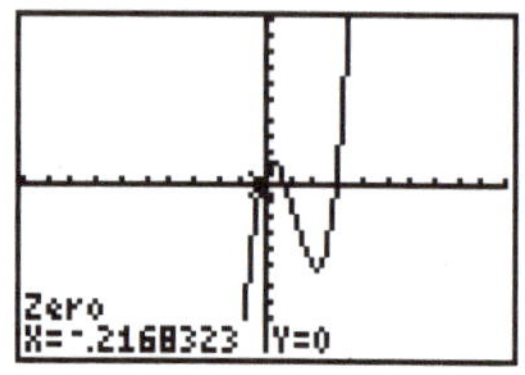

The screen displays $x = -.2168323$ as an approximation to the zero.
Proceeding as above for the other two zeros yields the approximations 0.78707062 and 2.9297617.

The procedures for finding local minima and maxima are the same as above except we press:
2nd [CALC] **3** and **2nd** [CALC] **4** instead of **2nd** [CALC] **2**.

This procedure yields a local minimum at (2.0946…, -5.048…) and a local maximum at (0.2387…, 1.344…).

WORKED EXAMPLE 3

Investigate how the graph of the function $f(x) = \alpha x^2$ changes as the value of α changes.

Solution

To study how the graph of $f(x)$ changes as α changes, we graph the family of functions:

$f(x) = 0.5x^2,\quad f(x) = x^2,\quad f(x) = 2x^2$

$f(x) = -0.5x^2,\quad f(x) = -x^2,\quad f(x) = -2x^2$

How can we graph a set of functions quickly?

To graph these functions we could define them as Y_1, Y_2, etc. however, we can define Y_1 as the set of functions:

$$\{0.5x^2, x^2, 2x^2, -0.5x^2, -x^2, -2x^2\}$$

Press **Y =** then enter $\{0.5x^2, x^2, 2x^2, -0.5x^2, -x^2, -2x^2\}$

Remember: Use the **(–)** key and *not* the subtraction key to enter the negative coefficients.

We obtain the graphs shown in the display.

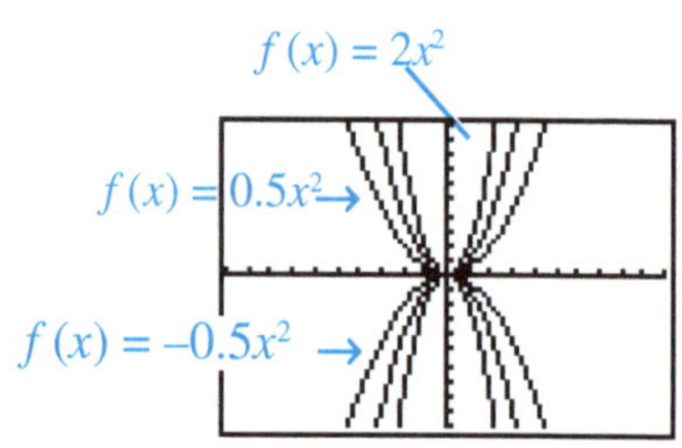

We observe that as $|\alpha|$ increases, the graph closes. The graph opens upward for $\alpha > 0$ and downward for $\alpha < 0$.

Exercises

1. Define a function that satisfies Dirichlet's but not Euler's definition of a function.

2. Does there exist a function that satisfies Euler's but not Dirichlet's definition? Explain your answer.

3. Which of the following figures is the graph of a function as defined by Cantor? Justify your answers.

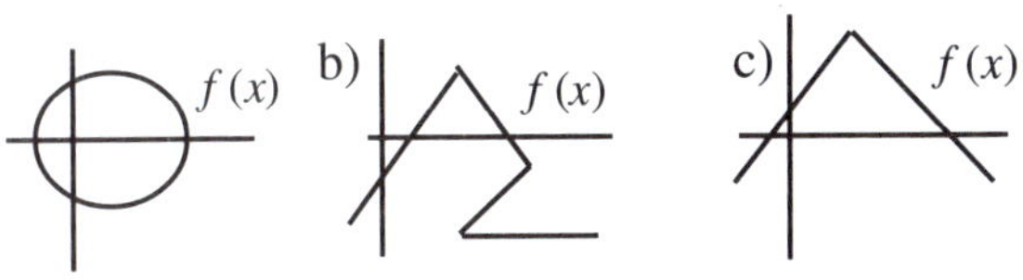

4. A vertical line through a graph intersects it in two points. Does the graph define a function? Explain.

5. State the domain and range of $f(x)$.

a) $f(x)=3x-7$ b) $f(x)=5x^2-9$ c) $f(x)=x^3-4$

d) $f(x)=\sqrt{x-3}$ e) $f(x)=\sqrt{x^2-16}$ f) $f(x)=\sqrt[3]{x}$

6. Use your graphing calculator to find the approximate range and the extrema of $f(x)$.

a) $f(x) = x^2 + 4x - 21$ b) $f(x) = 7x - x^2 - 19$

c) $f(x) = x^5 + x - 13$ d) $f(x) = x^4 - 9x + 5$

Note: To restore the range values of your calculator to $-10 \le x \le 10$, $-10 \le y \le 10$, press these keys:

ZOOM **6** **ENTER**

7. A function of the form $f(x) = ax^2 + bx + c$ where a, b, and c are constants is called a *quadratic function*. The graph of a quadratic function is called a *parabola*. Every parabola is symmetric about a vertical line called its *axis of symmetry*. The axis of symmetry passes through a point on the parabola called the *vertex*. The figure shows the parabola corresponding to the quadratic function $f(x) = x^2 - 10x + 21$.

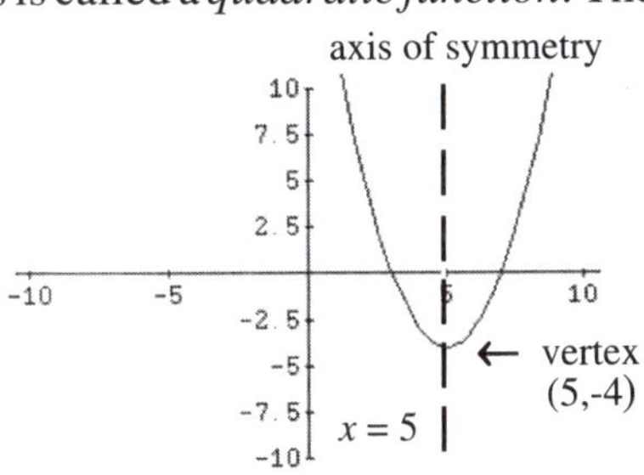

Find the coordinates (to 2 decimal places) of the vertex and the equation of the axis of symmetry of the graph of $f(x)$.

a) $f(x)=x^2-2x-35$ b) $f(x)=17-19x-x^2$

Investigations

8. We propose to investigate the changes in the graph of the quadratic function $f(x) = ax^2 + bx + c$ as we vary the coefficients, a, b, and c.

a) To investigate the changes resulting from varying c, graph the quadratic functions:

$$x^2+x+3,\quad x^2+x,\quad x^2+x-3$$

Write the equation of the axis of symmetry and the coordinates of the vertex of each parabola? Describe how the parabola changes as c changes.

b) To investigate the changes resulting from varying b, graph the quadratic functions:

$$2x^2-2x,\quad 2x^2-4x \quad\text{and}\quad 2x^2-8x$$

Write the equation of the axis of symmetry and the coordinates of the vertex of each parabola? Is there a relationship between the equation of the axis of symmetry and the value of b? To test your conclusions, graph the quadratic functions, x^2-2x, x^2-4x and x^2-8x. Answer the questions above for these functions and look for a pattern. How can you find the axis of symmetry of the parabola from the values of a and b?

c) Use the formula you have discovered above, to write an expression in terms of a and b for the coordinates of the vertex of a parabola (defined by $f(x) = ax^2 + bx + c$).

CHALLENGE

The trajectory of a baseball after it leaves the bat can be described by the equation $h(x) = -0.05x^2 + 5.4x$ where $h(x)$ denotes the height of the ball when it has traveled x yards from home plate.

Use your graphing calculator to approximate:

- the greatest height reached by the baseball
- the horizontal distance of the ball from home plate when it reaches its greatest height.
- the horizontal distance traveled by the ball

Can you use your results from *Exploration* **8** to calculate exact answers to these questions?

EXPLORATION 8 Maxima & Minima of Polynomials

Who Hit the Longest Home Run in Major League History?

THE BETTMANN ARCHIVE

Mickey Mantle 1931-1995

On September 10, 1960, Mickey Mantle hit the longest home run ever recorded in regular-season major league baseball. In a game between the New York Yankees and the Detroit Tigers at Briggs Stadium in Detroit, he sent the ball into a parabolic orbit. The trajectory of the ball is given by the equation

$$y = 0.9x - 0.0014x^2$$

where x represents the horizontal distance (in feet) and y the vertical distance (in feet) of the ball from home plate. During his career with the New York Yankees (1951-68), Mantle hit a total of 18 home runs in World Series play. This is still a major league record.

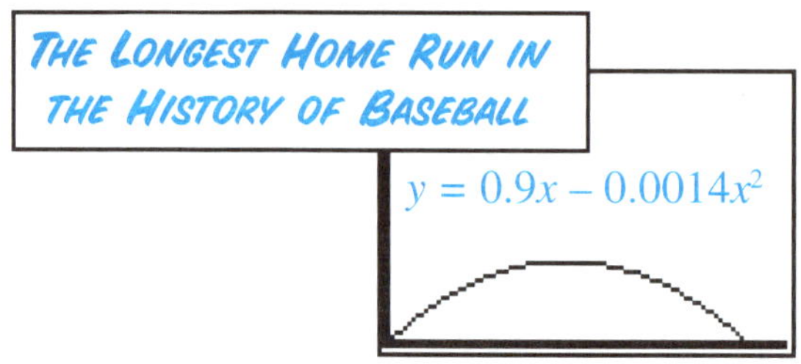

Worked Example

a) Graph the trajectory of baseball's longest home run.
b) Determine the maximum height reached by the ball.
c) Determine how far the ball landed from home plate.

Solution

a) We could begin by graphing the curve defined by $y = 0.9x - 0.0014x^2$. However, in the default window, we would see only the first part of the trajectory and it would be necessary to press ZOOM 3 ENTER three times to see the entire curve.

Try this and see what happens.

Alternatively, we observe that $0 \le x \le 700$, (choosing 700 feet as an estimate of the upper limit to the distance the ball will travel). Similarly we consider the height of the ball to occur in the interval, $0 \le y \le 460$. To set this window, we press WINDOW and enter these settings. We then enter the equation and graph it.

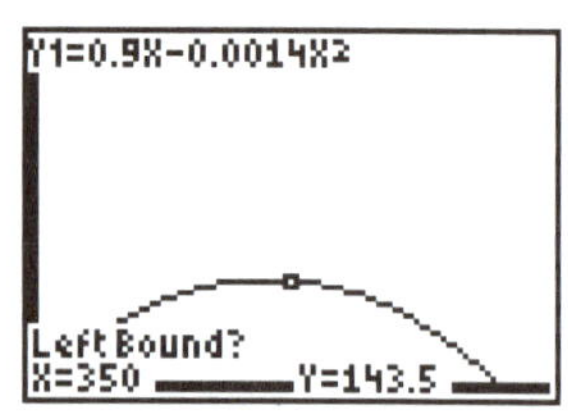

b) To determine the coordinates of the maximum height, we press: 2nd [CALC] 4 . In response to the prompt **Left Bound?**, we trace along the curve using the cursor key, to a point left of the maximum and press ENTER . In response to the prompt **Right Bound?** we trace to a point right of the maximum and press ENTER .
Pressing ENTER one more time, yields the display shown.

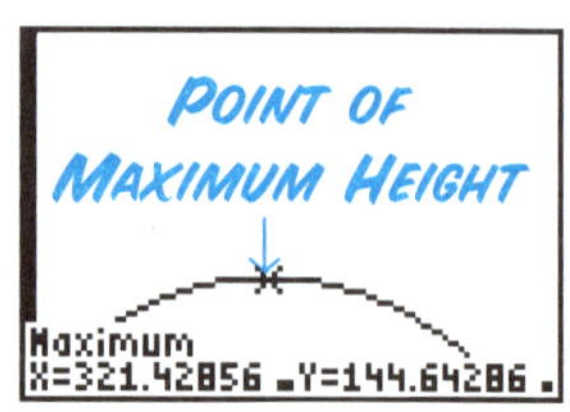

c) That is, the maximum height of the ball was about 145 ft . Since the maximum is the vertex of the parabola, it is on the line of symmetry and therefore mid way between the zeros. The length of the longest home run is: 2×321.428 feet or 643 feet (to the nearest foot).

Exercises

1. Write the domain and range of the function in the *worked example* which describes the trajectory of the longest home run in baseball history.

2. In the *worked example*:

a) What is the name of the curve that describes the trajectory of the baseball?

b) What is the horizontal distance of the ball from home plate when it reaches the highest point in its trajectory?

c) What is the slope of the secant which joins the point on the trajectory at $x = 0$ to the highest point in the trajectory?

d) What is the slope of the tangent to the trajectory at its highest point? Give reasons for your answer.

e) The equation of the trajectory of Mickey Mantle's home run is a mathematical model of the path of the baseball. Do you think this model is accurate? Explain.

3. A function has a *local maximum at* $x = a$ if and only if $f(x) \leq f(a)$ for all values of x close to a. The value $f(a)$ is said to be a *local maximum*.

a) Define a *local minimum*.

b) Does the function defined by the equation $y = x^3 + 8$, have either a local maximum or a local minimum?

4. Graph the function $f(x) = 2x^3 - 13x^2 - 7x + 1$. Use the CALC menu to determine all local maxima and minima.

5. Determine all the points at which each of these functions has a local maximum or minimum.

a) $f(x) = 3x^2 - 7x + 5$

b) $f(x) = x^4 + 2x^3 - 3x^2 - 4x - 4$

c) $f(x) = x^4 - 6x^3 + 12x^2 - 8x - 5$

6. A local maximum or minimum is called an *extremum*. What is the largest number of extrema which $f(x)$ can have if $f(x)$ is:

a) a linear function? b) a quadratic function?

c) a cubic function? d) a polynomial of degree n?

7. Write an equation to define a function of x which has a maximum when $x = 0$, and a minimum when $x = 4$. Is the function you have defined the only one satisfying these conditions?

Investigations

8. At the Summer Olympics in August 1996, Michael Johnson set a new record by running the 200-m dash in 19.32 seconds. His distance s (in meters) as a function of time t (in seconds) can be closely approximated by the cubic function.

$$s(t) = -0.0419t^3 + 1.239t^2 + 2.048t + 0.443$$

a) Graph Michael Johnson's distance as a function of time, where the distance is in meters and the time in seconds.

b) Create a table to determine the distance traveled by Johnson at these times:

(i) 5 seconds (ii) 10 seconds (iii) 15 seconds

c) According to this mathematical model, how many seconds did it take Michael Johnson to run:

- the first 100 meters?
- the second 100 meters?
- the entire 200-*m* race?

Hint: Graph the line $y = 100$ and use the CALC menu to find the point at which it intersects Johnson's distance-time graph.

d) Use your table or the graph to estimate the time when Michael Johnson reached his maximum speed in the race.

CHALLENGE

WHAT IS THE MAXIMUM POSSIBLE AREA?

Two sides of a triangular sign are to be each 5 m long. What length of the third side would yield the maximum possible area? (Give your answer to the nearest centimeter.)

EXPLORATION 9 Polynomials to Rational Functions

©1996 by Sidney Harris

"We have reason to believe Bingleman is an irrational number himself."

A *polynomial function* is any function of the form:

$$f(x)=a_n x^n + a_{n-1}x^{n-1} + \ldots a_1 + a_0$$

where a_i is a real number (for all i) and n is a positive integer.

We leave as an exercise the proof that the sum, difference and product of any pair of polynomial functions are also polynomial functions. However, the **quotient** of two polynomial functions is not necessarily a polynomial function.

Since applications often yield expressions that are quotients of polynomials, mathematicians defined a *rational function* to be any function of the form:

$$f(x)=\frac{g(x)}{h(x)}$$

where $g(x)$ and $h(x)$ are polynomial functions. The worked examples show the graphs of some rational functions. The cartoon shows the malfunction of an irrational number.

It follows readily from the definition that the domain of a polynomial function is always the set of all real numbers. That is, a polynomial function $f(x)$ has a value for every (finite) value of x. However, a rational function is defined for all real values x **except** those for which the denominator is zero. A rational function is said to have a *singularity* at each point at which its denominator is zero. Hence the domain of a rational function is the set of all real numbers except for those at which the rational function has a singularity.

WORKED EXAMPLE 1

Graph the function $f(x)=\frac{x}{x-3}$ and state its domain and range.

Solution

To graph the function, press the following keys:

Y = | X,T,θ,n | ÷ | (| X,T,θ,n | – | 3 | ENTER | GRAPH

Note: We use brackets to indicate that x is to be divided by the entire expression $x - 3$. The final bracket may be omitted.

We observe from the graph that:

the domain of $f(x)$ is: $\{x \mid x \text{ is real and } x \neq 3\}$
the range of $f(x)$ is: $\{y \mid y \text{ is real and } y \neq 1\}$

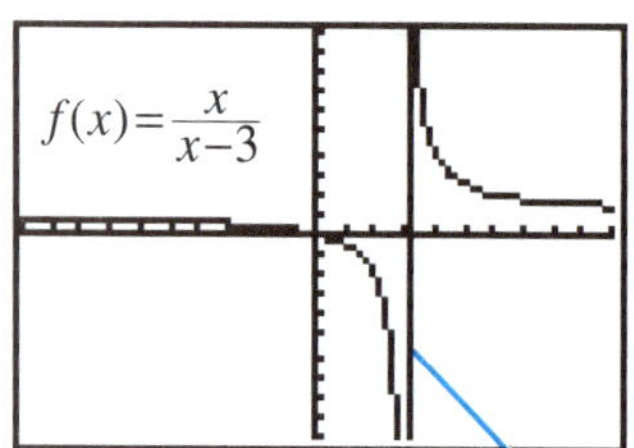

The line with equation $x = 3$ is not part of the graph

WORKED EXAMPLE 2

Graph each function and state its domain, range, zeros and extrema.

a) $f(x)=\dfrac{x}{x^2-x-12}$ b) $f(x)=\dfrac{8x+1}{x^2+1}$

What is the behavior of each graph as $x \rightarrow \infty$?

Solution

a) To plot this graph we use the following keying sequence.

Y = **X,T,θ,n** **÷** **(** **X,T,θ,n** **x^2** **–** **X,T,θ,n** **–** **1** **2** **ENTER** **GRAPH**

This sequence generates the graph in the display.

Tracing along the graph, we discover singularities at $x = -3$ and 4:

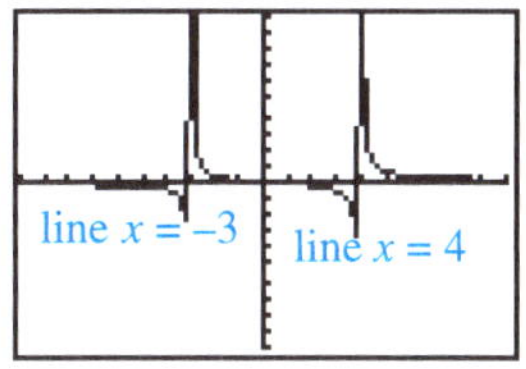

domain of $f(x)$: $\{x \mid x$ is real and $x \neq -4$ or $3\}$

range of $f(x)$: the set of all real numbers

zeros of $f(x)$: The fraction is zero when the numerator is zero; that is, when $x = 0$. The graph suggests that $f(x)$ approaches zero asymptotically as $x \rightarrow \infty$ and as $x \rightarrow -\infty$

> **Note:**
> To approximate the values at which the singularities occur, we use the **TRACE** key.

extrema of $f(x)$: The function $f(x)$ has no extrema; i.e. no relative maxima or minima. However, the graph suggests $f(x) \rightarrow -\infty$ as $x \rightarrow -3$ or $x \rightarrow 4$ from the left and $f(x) \rightarrow \infty$ as $x \rightarrow -3$ or $x \rightarrow 4$ from the right. That is, the absolute value of the function increases without limit at the singularities at $x = -3$ and $x = 4$. (We can verify singularities at -3 and 4 by substitution.)

b) To plot this graph we use the following keying sequence.

Y = **(** **8** **X,T,θ,n** **+** **1** **)** **÷** **(** **X,T,θ,n** **x^2** **+** **1** **ENTER** **GRAPH**

This sequence generates the graph in the display.

The graph suggests:

domain of $f(x)$: the set of all real numbers

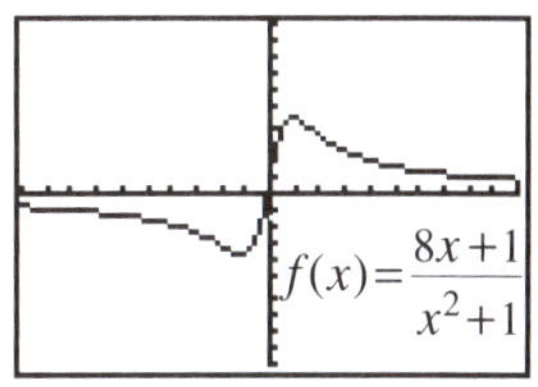

zeros of $f(x)$: The fraction is zero when the numerator is zero; that is, when $x = -\ 0.125$.
The graph suggests that $f(x)$ approaches zero asymptotically as $x \rightarrow \infty$ and as $x \rightarrow -\infty$.

extrema of $f(x)$: We press: **2nd** [CALC] **3** and **2nd** [CALC] **4** respectively as in *worked example* 2 (p. 34) to find a minimum at (-1.132…, -3.531…) and a maximum at (0.882…, 4.531…)

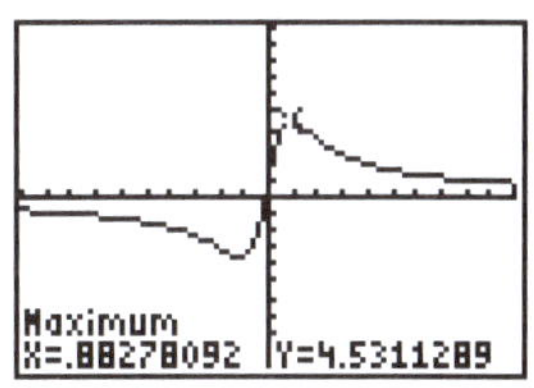

range of $f(x)$: The values of y vary between the minimum and maximum values of $f(x)$. The range is $\{y \mid -3.531\ldots \leq y \leq 4.531\ldots\}$.

WORKED EXAMPLE 3

For $f(x)$, state the domain, range, zeros and extrema to four decimal places where $f(x)$ is defined by.

$$f(x) = \frac{8x}{x^2 - x - 4}$$

What is the behavior of $f(x)$ as $x \to \infty$?

Solution

In *worked example* 2, we graphed functions and traced along each curve to find the approximate domain, range, zeros and extrema. When we need precise values for the features of a rational function, we proceed as follows:

To graph $f(x)$, we use the following keying sequence.

[Y=] [8] [X,T,θ,n] [÷] [(] [X,T,θ,n] [x^2] [–] [X,T,θ,n] [–] [4] [ENTER] [GRAPH]

This sequence generates the graph in the display.
It appears that there are two singularities, but *no extrema.*

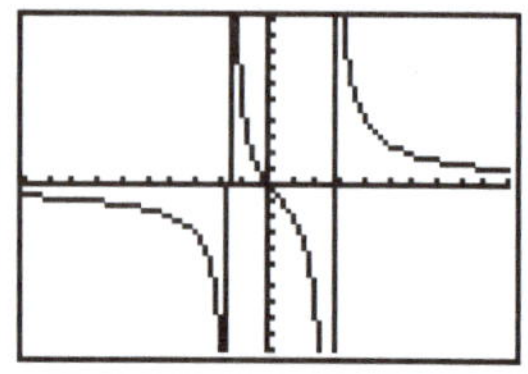

When we trace along the graph to the left, we reach the point, (-1.489362, 40.743207). Pressing [◄] one more time moves the cursor to the point (-1.702128, -22.71903). Since the y-coordinate changes sign in the interval $-1.702128 \le x \le -1.489362$, we check for a zero in this interval by pressing [2nd] [CALC] [2].

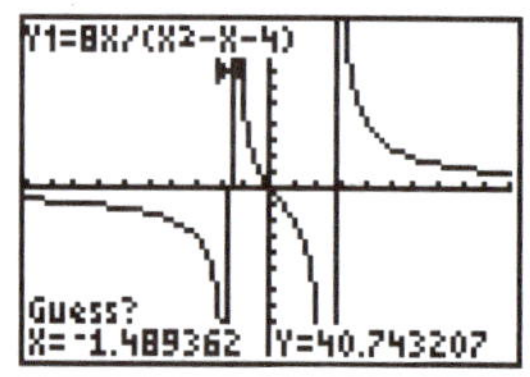

•In response to the prompt **Left Bound?**, press: –1.702128 [ENTER].

•In response to the prompt **Right Bound?**, press: –1.489362 [ENTER].

•In response to the prompt **Guess?**, press: [ENTER].

The screen displays the message: **ERR: SINGULARITY**, indicating that there is a singularity in this interval. To find the position of this singularity to 4 decimal places, we must find the value of x for which the denominator is 0. That is, we must solve the equation, $x^2 - x - 4 = 0$. To solve this equation, we graph the corresponding parabola $y = x^2 - x - 4$ and observe that this parabola has two zeros. To calculate the corresponding roots, we trace along the parabola to each root in turn and press:

[2nd] [CALC] [2]

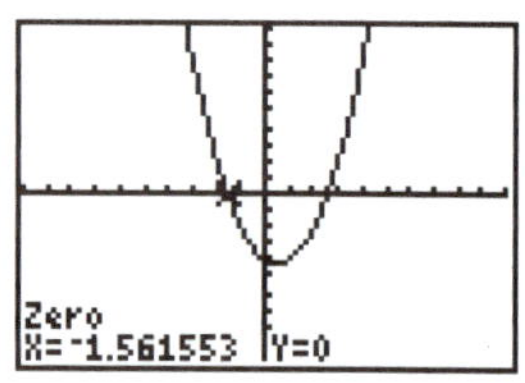

We obtain the roots, $x = -1.561553\ldots$ and $x = 2.561552$. How do we know that these roots are correct to four decimal places?

To test x = -1.561553, we press [2nd] [CALC] [1] and evaluate the expression, $y = x^2 - x - 4$ at x = -1.561554.
We obtain y = 4.8949E-6. Similarly we evaluate at x = -1.561552 and obtain y = -3.351E-6. Since the value of y changes sign in the interval, $-1.56164 \le x \le -1.56155$, then $x = -1.5616$ is a root correct to within 4 decimal places.

Note:
We could have solved $x^2 - x - 4 = 0$ using the quadratic formula, however the method used here will work with a polynomial of any degree. (See exercise 13.)

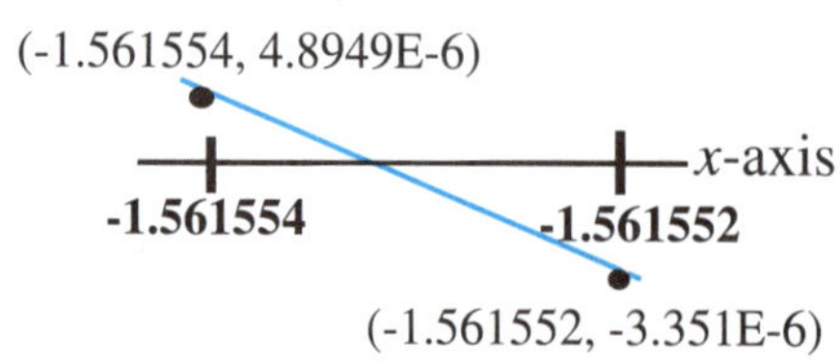

Similarly, we find the other root is x = 2.5616 to 4 decimal digits.

Therefore: domain of $f(x)$: $\{x \mid x \text{ is real and } x \ne -1.5616 \text{ or } 2.5616\}$
range of $f(x)$: the set of all real numbers
zeros of $f(x)$: $x = 0$ and approaches 0 as $x \to \pm\infty$.

Exercises

1. Define a rational number. How are the definitions of rational numbers and rational functions alike?

2. Is a polynomial a rational function? Explain.

3. What is a *singularity* ? Explain how you would find the singularities of a rational function.

4. Prove that the sum, difference, product and quotient of any pair of rational functions is a rational function.

5. Find the singularities of each of the following rational functions without using your calculator.

a) $f(x)=\dfrac{1}{x}$ b) $f(x)=\dfrac{(x-3)}{x^2-5x+6}$ c) $f(x)=\dfrac{x^3}{x^2-3x+7}$

Describe the behavior of each function as $x \to -\infty$ and as $x \to \infty$.

6. Graph the functions in *exercise* 5 to check your answers. Does a rational function always approach infinity near a value of x for which the denominator is zero? Explain (See *exercise* 5.)

7. Write an algebraic expression for a rational function whose domain is the set of all the real numbers except 4 and 7 and whose range is the set of all real numbers except 0.

8. Graph the functions $f(x) = \dfrac{1}{x-5}$ and $f(x) = \dfrac{1}{x} - 5$. In what way do the keying sequences for graphing these two functions differ? Name the singularities of these two functions. Describe the behavior of each function as $x \to -\infty$ and as $x \to \infty$.

9. Write keying sequences which use the x^{-1} key to graph the functions in *exercise* 8. Compare with sequences which do not use this key. Which method uses fewer keystrokes?

10. Graph each function. Use TRACE to determine the domain, range and extrema of each function to two decimal places.

a) $\dfrac{x}{9-7x+x^2}$ b) $\dfrac{7x+3}{x^2-2}$ c) $\dfrac{x^3}{x^2-x-12}$

Investigations

11. The function $f(x)$ in worked example 1 gets closer and closer to the value 1 as x approaches ∞. Graph $f(x)$ and then press TRACE . Hold down the ▶ key and watch the cursor move along the curve to the right. Observe how it jumps the singularity at $x = 4$ and traces along the right branch of the curve. Continue to hold down the key until the cursor reaches the point with x-coordinate 100. What is the corresponding y-coordinate? If you trace far enough to the right, will the x- coordinate eventually reach the value 1? Justify your answer algebraically. Calculate the value of $f(x)$ corresponding to $x = 1000$.

When the graph of a function $f(x)$ approaches the line $y = \alpha$, so that $f(x) - \alpha \to 0$, as x becomes infinite, we say $f(x)$ approaches α *asymptotically* as x approaches infinity. In the example above, $f(x)$ approaches the value 1 asymptotically.
Explain what happens to $f(x)$ as x approaches $-\infty$.

12. Graph the function $f(x)$ defined by:

$$f(x) = \frac{5x^2}{2x^2-x-12}$$

a) Use the trace function to determine what value(s) $f(x)$ approaches as $x \to -\infty$ and as $x \to \infty$.
b) Find the singularities and extrema of $f(x)$ correct to 4 decimal places.
c) Write the domain and range of $f(x)$.

13. Graph the function $f(x)$ defined by:

$$f(x)=\frac{4x^2-5}{2x^3-3x^2+4x-6}$$

a) Use the trace function to determine what value(s) $f(x)$ approaches as $x \to -\infty$ and as $x \to \infty$.
b) Find the singularities and extrema of $f(x)$ correct to 4 decimal places.
c) Write the domain and range of $f(x)$.

14. Construct a rational function which has singularities at $x = -1.5$, $x = 2$ and $x = 5$; and which approaches the line $y = 3$ asymptotically as $x \to -\infty$ and as $x \to \infty$.

15. Express each function as a sum or difference of partial fractions. Then determine the domain, range, singularities, zeros and extrema of each function.

a) $\dfrac{2}{x^2-4x+3}$ b) $\dfrac{x^2+3x-21}{x^3-7x^2}$ c) $\dfrac{x^2-x-3}{x^3-6x^2+9x-4}$

Use your TI-84 Plus to verify your answers.

EXPLORATION 10 Optimization Problems

The recent realization that the world's natural resources are in limited supply has made conservation an international priority in this century. Aluminum, one of the world's non-renewable resources, is used in the production of about 100 billion beverage cans per year. Two thirds of these cans are recycled. The worked example below is typical of many types of problems that applied mathematicians and engineers confront in the design of containers to preserve resources and optimize costs.

WORKED EXAMPLE

An aluminum beverage can must be designed to hold 355 mL of soda. What dimensions for the can would require the minimum amount of aluminum? (Give the dimensions to the nearest hundredth of a centimeter.)

Solution

We assume the can is to be a right circular cylinder (for stacking purposes). Denote its height and the radius of its base (in centimeters) by **h** and **r** respectively.

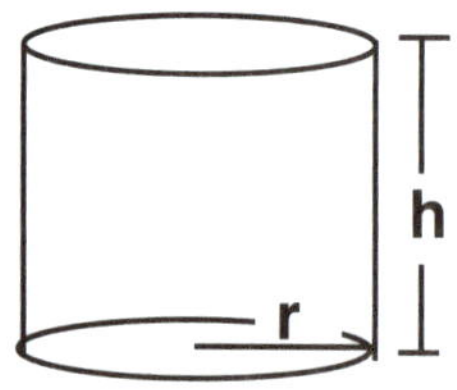

We must find the values of **r** and **h** which minimize the surface area **S**, while maintaining a volume of 355 cm^3 (Recall $1\ \text{cm}^3 = 1$ mL). That is, we must minimize **S** subject to the condition $\pi r^2 h = 300$ where:

$$S = 2\pi r^2 + 2\pi r h$$

$2\pi r^2$ ⇑ Total Area of top and bottom; $2\pi r h$ ⇑ Area of curved surface

Since $\pi r^2 h = 355$, we can substitute $\dfrac{355}{\pi r^2}$ for **h** in the formula for **S** to obtain:

$$S = 2\pi r^2 + \frac{710}{r}$$

To graph $y = 2\pi x^2 + \dfrac{710}{x}$ we press these keys:

Y= 2 2nd [π] X,T,θ,n x^2 + 7 1 0 ÷ X,T,θ,n ENTER GRAPH

In the default range $-10 \le x \le 10$; $-10 \le y \le 10$, the display shows only the steep segment of the graph where it crosses the x-axis at $x = -4.83\ldots$. To determine an appropriate viewing window for the graph near the minimum, we press **2nd** [TABLE] and obtain the table of values shown in the display. This table gives the values of the function corresponding to $x = 1$ through 6. By pressing the **▼** key, we can view the table for greater values of x.

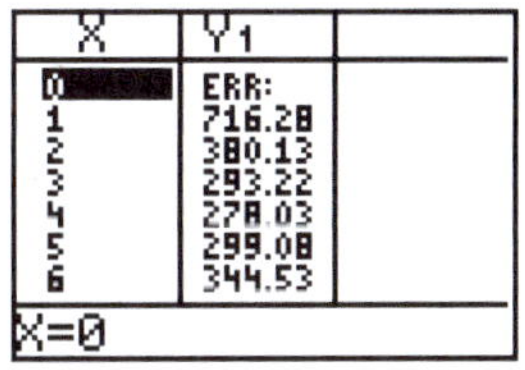

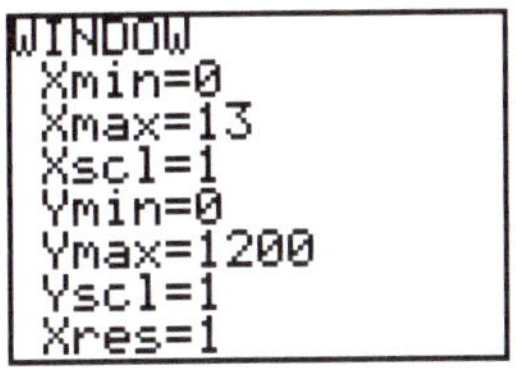

The table reveals that there is a singularity at $x = 0$, and as x increases, the value of the function appears to decrease until $x = 4$ where the function takes a value of 278.03. Somewhere in this vicinity, the function has a minimum and begins to increase, reaching the value of 1116.5 at $x = 13$. To graph the curve in this interval, we press **WINDOW** and enter the settings shown in the display.

Upon pressing **GRAPH**, we obtain the graph shown in the display. To determine the coordinates of the local minimum, we press:

2nd [CALC] **3**

- In response to the prompt **Left Bound?**, press: **0** **ENTER**.
- In response to the prompt **Right Bound?**, press: **13** **ENTER**.
- In response to the prompt **Guess?**, press: **ENTER**.

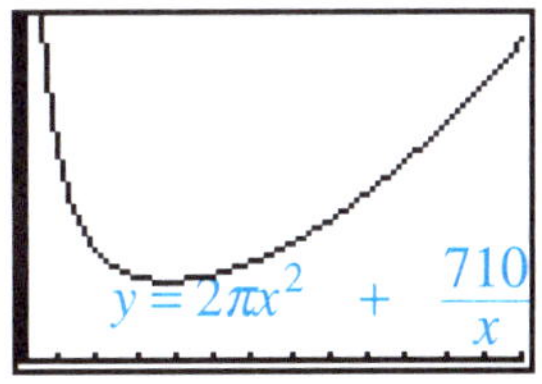

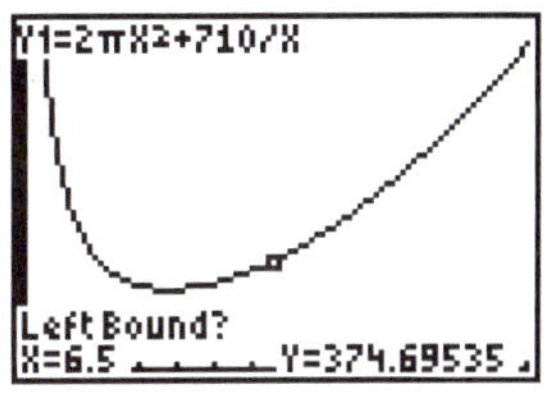

The display shows that the coordinates of the local minimum are (3.8372138, 277.54502). This means that, the amount of aluminum used will be a minimum when the radius is about 3.84 cm and its height is $355/(\pi \times 3.84^2)$ or 7.66 cm. With these dimensions the surface area will be approximately 278 cm^2.

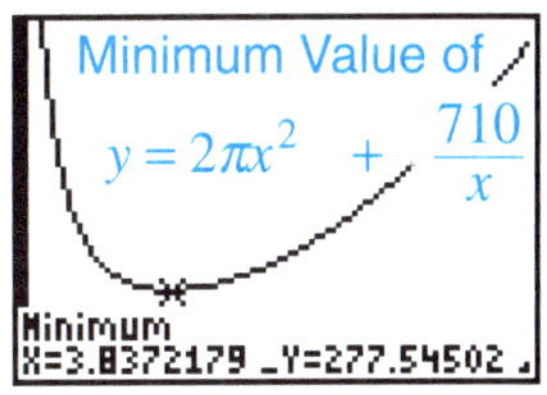

You may have noticed that a cylindrical can of fixed volume has minimum surface area when its height is equal to its diameter. Is the height of an aluminum beverage tin equal to its diameter? If not, write a sentence to explain why you think beverage tins are made with a height-to-diameter ratio which is not optimal.

Problems like the one above that seek the maximum or minimum value of a function subject to a constraint are called **optimization problems.**

Exercises

1. What is an optimization problem?
In the worked example:
a) what is the function for which we seek a minimum value?
b) what equation defines the constraint?

2. A wire of length 3.6 m is to be bent into a rectangular shape. Determine the length and width of the rectangle to enclose a maximum area.

3. Determine (two decimal places) the coordinates of the point on the line with equation $y = 3x + 2$ which is closest to the origin, (0, 0).

4. What point on the parabola defined by equation $y = x^2 - 4x + 3$ is closest to the origin?

5. Determine distance between the lines defined by the equations $y = 2.7x - 13$ and $y = 2.7x + 16$. (Give your answer correct to two decimal places.)

6. Answer *exercise* 5 by finding the point of intersection of the line defined by $y = 2.7x + 16$ and the line perpendicular to it and passing through the point (0, -13). Then compute the distance between the point of intersection and point (0, -13).

7. In a heavy fog at 7:00 a.m., a cargo ship is 80 km due east of a luxury liner which is sailing due south at a speed of 42 km/h. If the cargo ship is sailing due west at 48 km/h, at what time will they attain their closest approach? How close will they be?

8. The total cost in dollars of producing x grummets is given by the formula $C = 0.35x^2 + 23x + 20$. Each grummet sells for \$80.
a) How many grummets should be produced to achieve a maximum profit?
b) How much profit would this generate?

9. Ms. Chiu plans to put a rectangular swimming pool in her yard. Since one side of the pool will run along an existing fence it will be necessary for her to fence only three sides. What dimensions of the enclosed rectangle would maximize the contained area if Ms. Chiu has 35 m of fencing?

Investigations

10. The diagram below shows a beam of length L laid across a wall 7 m tall to support the side of a building. The wall is 11 m from the building.

a) Denote the distance between the foot of the beam and the wall by x. Write an equation which relates x and h where h is the height of the upper end of the beam.
(Use similar triangles.)

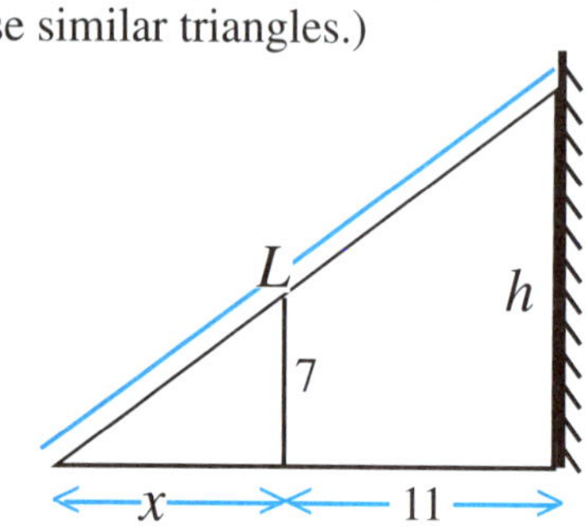

b) Write an equation for the length of the beam L in terms of x and h.
c) Use your equations in parts a and b of this exercise to express L in terms of the variable x only.
d) Graph L as a function of x and determine the minimum possible length of the beam.
e) How far is the bottom of the beam from the wall when the beam is the minimum possible length?
f) Does the function of x you graphed have more than one minimum? Intepret each minimum?
g) Is there a maximum length which the beam can have? Explain your answer.

CHALLENGE

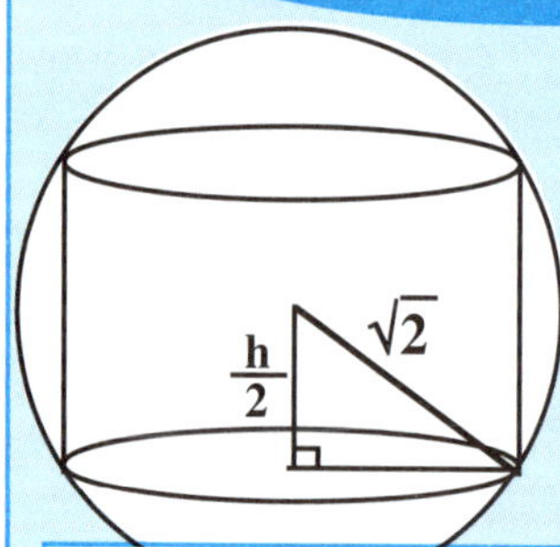

Determine the height of a right circular cylinder of maximum volume which can be inscribed in a sphere of radius $\sqrt{2}$

Part II

Elementary SPECIAL FUNCTIONS A Short History

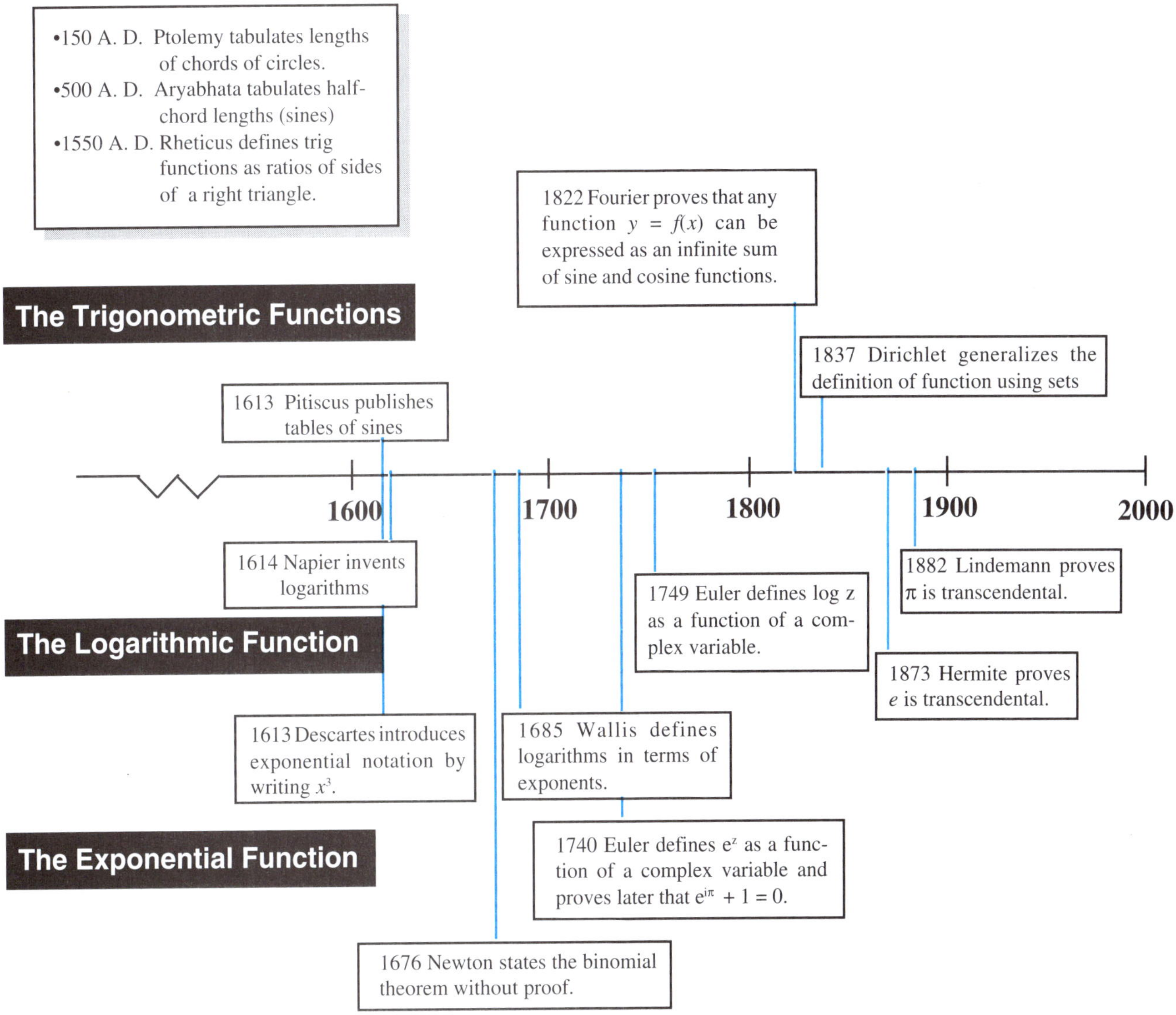

In *Exploration 1*, we observed that the earliest functions were defined as tables of values. With the development of algebra during the Renaissance, functions were described as algebraic expressions and, after Descartes, as graphs. Transcendental functions, i.e., those that cannot be defined using algebraic expressions, also had their beginnings as tables.

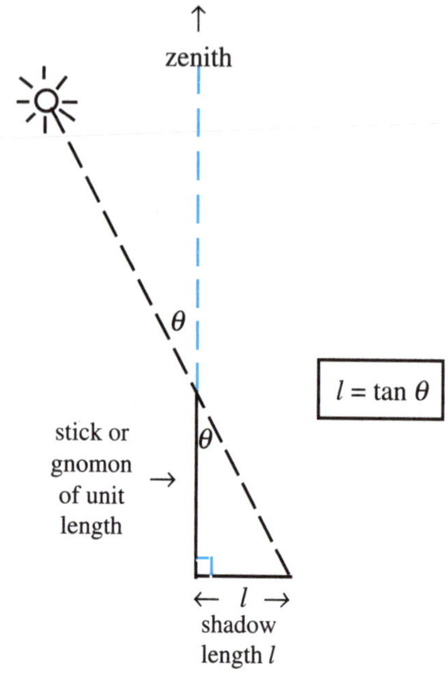

As early as the thirteenth century B.C., the ancient Egyptians made tables showing the lengths of shadows corresponding to the time of day. The diagram shows how the angle between the sun and the zenith is related to the length of the shadow of a vertical stick. This relationship, recorded at first as a table, was the precursor to the modern tangent function. The ancient Greek Eratosthenes (c. 276-194 B.C.) used the ratios of sides of triangles to calculate the circumference of the earth to within 99% of its true value. In 150 A.D., Ptolemy tabulated the lengths of chords of a unit circle, defining what are recognized today as the sine and cosine functions–the first transcendental functions to make their appearance.

Before the development of the scientific calculator in the late 1960s, scientists and engineers used tables to perform computations involving trigonometric functions. However, scientific calculators now use infinite series representations of the trig functions to display these function values.

solve vb. **To *solve a triangle* means to determine the measures of its 3 angles and its 3 sides.**

Trigonometry derives from the ancient Greek words *trigonon,* and *metros* meaning "triangle measure". Inspired by the study of astronomy and the development of navigation, trigonometry began as the quest to solve triangles. The ancient Greeks discovered that if they knew the lengths of two sides of a triangle and the contained angle, they could calculate the length of the third side. This was made possible by a theorem called the *Cosine Law*, found originally in Euclid's treatise, *The Elements.* The Cosine Law is a generalization of the Theorem of Pythagoras, as shown in the diagram below. We observe that when $\theta = 90°$, $\cos\theta = 0$ and the Cosine Law reduces to the Theorem of Pythagoras.

Cosine Law

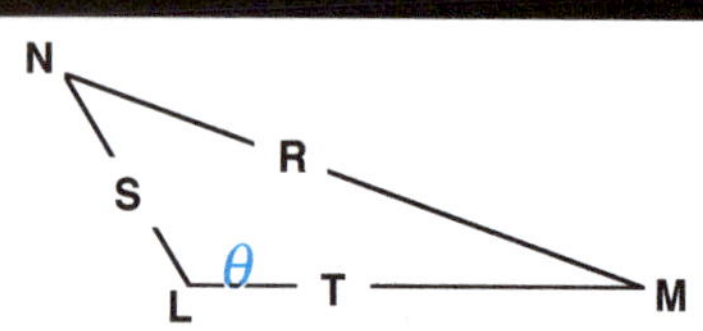

$$R^2 = S^2 + T^2 - 2ST\cos\theta$$

Theorem of Pythagoras

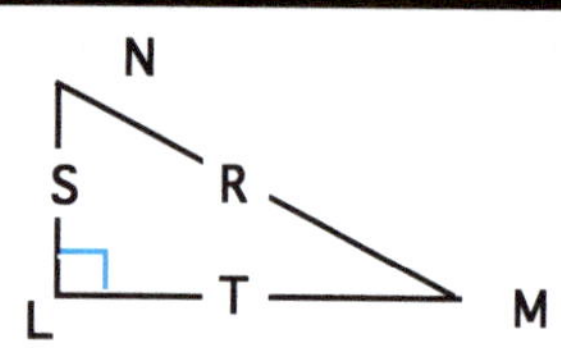

$$R^2 = S^2 + T^2$$

WORKED EXAMPLE 1

The par 4 fourteenth hole at the Glen Abbey G. C. is one of the toughest in golf. Golfers hitting from the gold tees must drive their tee shots a distance of about 258 yards across a river to a point S, on the fairway. From position S, the distance to the green is about 168 yards. What is the distance TG from the tee to the green as the crow flies if $\angle TSG$ is 149°?

Solution

Accessible from MODE

```
PROGRAM:SAS
:Fix 2
:Degree
:Input "ONE SIDE
?",S
:Input "OTHER SI
DE?",T
:Input "CONTAINE
D ANGLE?",A
:√(S²+T²-2STcos(
A)→R
:Disp R
:
```

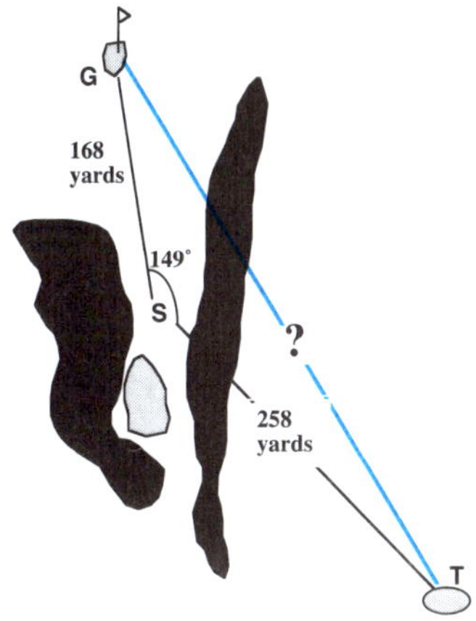

On solving for R in the Cosine Law, we obtain:

$$R = \sqrt{S^2 + T^2 - 2ST\cos\theta}$$

We could substitute the values of S, T and θ directly into this expression to calculate R, or if we have several such computations to perform, we merely formalize the computation in a program, SAS, (SIDE-ANGLE-SIDE) as shown in the display. When we run it, we obtain this display which indicates that TG is about 411 yards.

```
prgmSAS
ONE SIDE?258
OTHER SIDE?168
CONTAINED ANGLE?
149
          411.21
            Done
```

WORKED EXAMPLE 2

a) Write a program, ALLSIDES, to solve a triangle given the lengths of all three sides.
b) Use your program to solve the triangle with sides of length 831.7 m, 435.5 m and 514.4 m.

Solution

a) Program, ALLSIDES, shown in the display considers a general triangle LMN with sides of length R, S and T as shown. It requests input of the values R, S and T and then computes the angles using the cosine law.

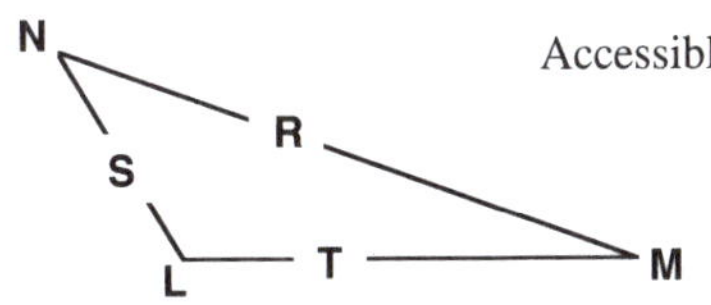

b) When we run ALLSIDES and input the lengths, 831.7, 435.5 and 514.4, in that order, we obtain this display which gives the angles of the triangle in degrees.

```
THIRD SIDE?514.4
ANGLES ARE
          122.00
           26.36
           31.63
            Done
```

Accessible from MODE

```
PROGRAM:ALLSIDES
:Fix 2
:Degree
:Input "THE LONG
EST SIDE?",R
:Input "SHORTEST
 SIDE?",S
:Input "THIRD SI
DE?",T
:If R≥S+T
:Then
:Disp "NO SUCH T
RIANGLE"
:Else
:If R²=S²+T²
:Then
:90→L
:Else
:cos⁻¹((R²-S²-T²)
/(-2S*T))→L
:End
:cos⁻¹((T²-R²-S²)
/(-2R*S))→N
:180-L-N→M
:Disp "ANGLES AR
E",L,M,N
:End
:Return
:
```

Observe the use of the If…Then…Else construction to isolate special cases.

Observe the use of the If…Then…Else construction to isolate special cases.

This is put here for later when we embed this program as a subroutine in a larger program.

Exercises

1. In the diagram below:

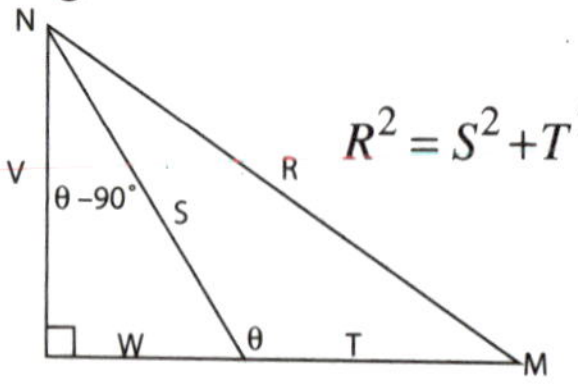

$$R^2 = S^2 + T^2 - 2ST\cos\theta$$

a) Use the Pythagorean Theorem to express R^2 in terms of V, W and T.
b) Express W in terms of S and θ.
c) Substitute from Part b) into your expression in Part a) to prove the Cosine Law.

2. a) Explain what is meant by the statement:
The Pythagorean Theorem is a special case of the Cosine Law.
b) What is the angle between equal sides of a triangle with two sides of length 60 *m* and a third side of length 80 *m*.

3. When it was completed around 2580 B.C., the Great Pyramid of Cheops was the tallest structure in the world. Its height was 147 m. Its base was a square with side length 230 m.

Calculate the angle of elevation of the top of the pyramid as measured by an observer located at a corner of the base.

4. In a "closest-to-the-hole" showdown, the golfer whose tee shot lands closest to the hole wins a new car. The first golfer hit a shot that stopped on the green 48 feet from the hole, exactly 212 yards from where he hit. The second golfer from the same position, hit a shot 203 yards to the green so that the hole was between the two golf balls and on the line joining them. If the angle between the positions of the two golf balls as seen from the tee was 8°, which golfer won the car?

5. Two roads diverge at an angle of 37°. Two motorcyclists leave the intersection at the same time along different roads. How far apart are they after 6 minutes if one cyclist has an average speed of 55 miles per hour and the other averages 48 miles per hour?

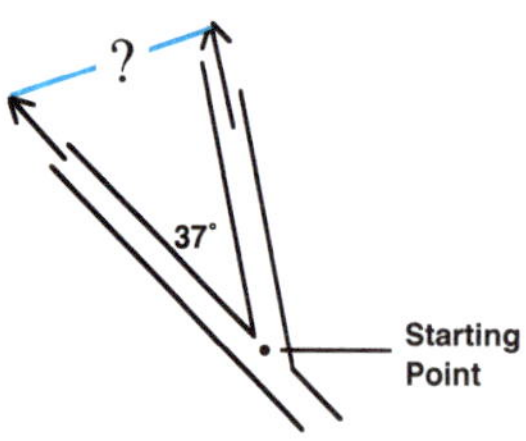

Investigations

6. Enter program ALLSIDES into your calculator. Run ALLSIDES to determine the measures of the angles of a triangular building lot with sides of length 45.2 m, 44.6 m and 38.6 m. Write the length of the side opposite each angle.

7. Run ALLSIDES to determine the measures of the angles of a triangle which has side lengths double those of the triangle in exercise 6. What do you discover? Explain.

8. Run ALLSIDES to verify that 12, 35 and 37 are the lengths of the sides of a right triangle.

9. The fourth hole at the Knotty Pine Golf and Country Club is a dog leg with dimensions as shown in the diagram. What is the distance from tee to green over the pond?

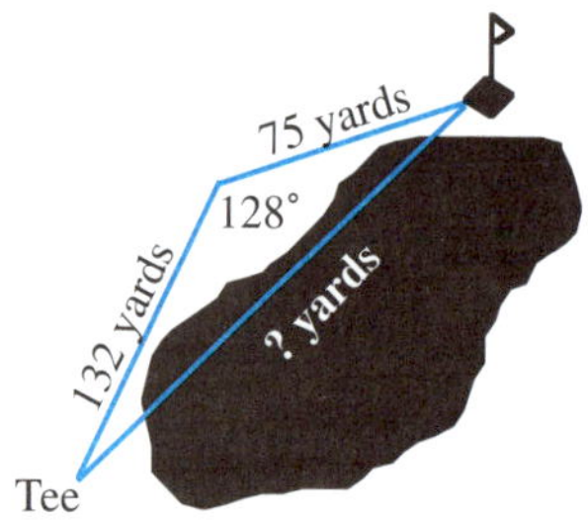

10. On October 1, 1976, the CN Tower, at a height of 553 m., became world's tallest free-standing structure.

CN Tower, Toronto, Canada

In the diagram below (not drawn to scale) ***A*** represents the farthest location on the earth's surface from which the tip of the Tower is visible. The distance along the surface of the earth from the base of the Tower to position ***A*** is denoted *s* and the angle subtended by *s* at the center of the earth is denoted by θ.

a) Calculate the angle θ in radians.

b) Use your answer in Part a) to calculate *s*.

c) On a clear day, what is the greatest distance from which the tip of the Tower is visible to an observer at sea level?

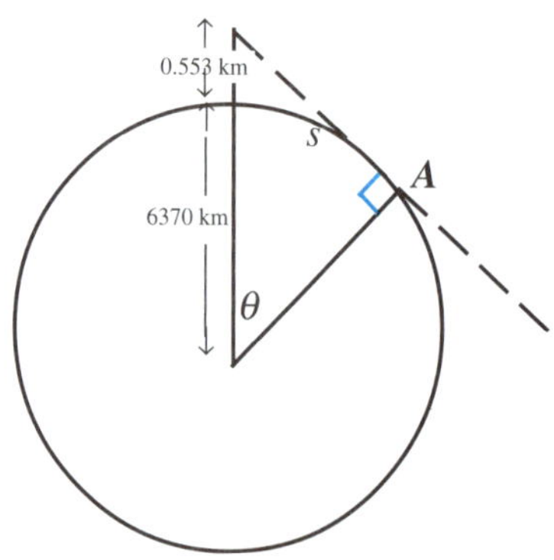

EXPLORATION 12 *Trigonometry & The Sine Law*

In this exploration, we shall discover how solving triangles enables us to calculate lengths and heights without traversing them and to calculate distances to stars which are entirely inaccessible. The Cosine law enables us to solve a triangle given the lengths of all its sides, or two sides and a contained angle. To solve all triangles we require one addition to the cosine law, called the *Sine law* or *Law of Sines*.

The Sine Law

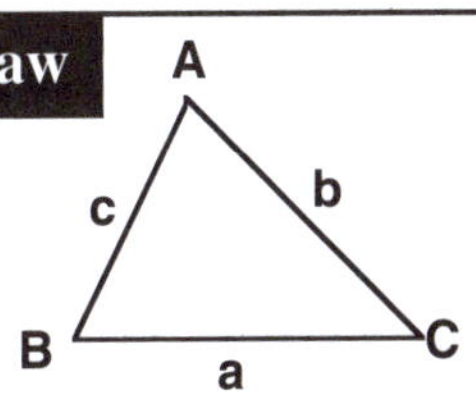

$$\frac{\sin A}{a} = \frac{\sin B}{b} = \frac{\sin C}{c}$$

How Far is the Nearest Star?

How do astronomers calculate the distance to the nearest star, Proxima Centauri?

WORKED EXAMPLE 1

The distance from the Earth to Proxima Centauri is determined by the parallax method. That is, the angle from earth to the star is measured when the earth is at point Y and found to be 83.903 224 5°. It is measured six months later when the earth is at point Z and observed to be 96.096 361 1°.

a) Write a program AAS (ANGLE-ANGLE-SIDE) which solves a triangle given the measures of two of its angles and any one of its sides.

b) Use program AAS to determine the approximate distance to Proxima Centauri.

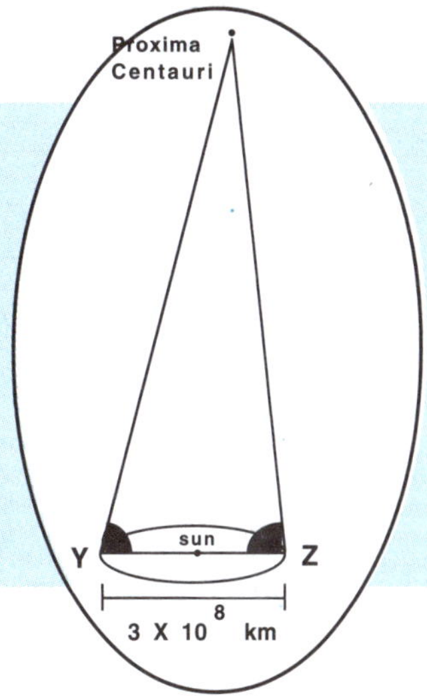

Solution

a) Using the given side and the opposite angle, we can calculate the ratio of the length of that side to the sine of the angle opposite. The Sine Law tells us that this ratio is the same for all sides, so we can use the known ratio to calculate the sides opposite the given angles. This calculation is executed in the program AAS shown in this display.

```
PROGRAM:AAS
:Float
:Degree
:Input "GIVEN SI
DE?",V
:Input "OPPOSITE
 ANGLE?",X
:Input "OTHER AN
GLE?",Y
:180-X-Y→Z
:Disp "ANGLES AR
E",X,Y,Z
:Pause
:Vsin(Z)/sin(X)→
W
:Vsin(Y)/sin(X)→
J
:Disp "SIDES ARE
",V,W,J
:Return
:
```

b) To determine the distance to Proxima Centauri, we calculate the angle at Proxima Centauri by subtracting the measures of the other two angles from 180°. The measure of the angle is 0.0004144°. We then enter in response to the prompts, the values 300000000, 0.0004144 and 83.9032245 in turn, as shown in the display.

```
0004144
OTHER ANGLE?83.9
032245
ANGLES ARE
         4.144E-4
       83.9032245
       96.0963611
```

The moving cursor in the upper right corner of the screen indicates that the calculator is in pause mode so you can view the screen. Pressing **ENTER** yields the second screen display giving the lengths of the sides, indicating that Proxima Centauri is about 41.2 trillion km or 4.4 light-years away.

```
       83.9032245
       96.0963611
SIDES ARE
        300000000
  4.124403053E13
  4.124399867E13
             Done
```

WORKED EXAMPLE 2

a) Write a program, SSA, to solve a triangle given the measures of two sides and an angle opposite one of the sides.

b) Run SSA to determine the measures of all the angles and sides of ΔABC if sides BC and AC are respectively 200 m and 160 m in length and $\angle ABC$ is 40°.

Solution

a) The diagram shows that if we are given an angle A, the length E, of one arm of angle A, and the length D, of the side opposite angle A, then there are 4 possible cases.

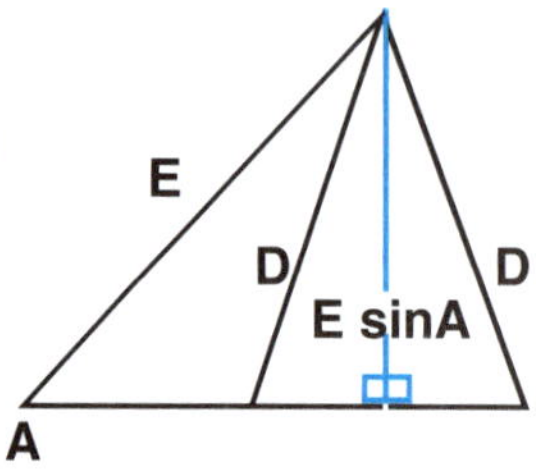

• $D < E\sin A$	no triangle possible
• $D = E\sin A$	unique right triangle
• $E\sin A < D < E$	two triangles
• $D \geq E$	unique triangle

When $E\sin A < D < E$, there are two triangles satisfying the given conditions. This is called the *ambiguous case*.

Program SSA uses branching to treat each case in turn.

```
PROGRAM:SSA
:Float
:Degree
:Input "GIVEN AN
GLE?",A
:Input "OPPOSITE
 SIDE?",D
:Input "OTHER SI
DE?",E
:If D<Esin(A)
:Then
:Disp "NO SUCH T
RIANGLE"
:Else
:sin⁻¹((E/D)sin(A
)→B
:180-B-A→C
:Dsin(C)/sin(A)→
F
:If D≥E or D=Esi
n(A)
:Then
:Disp "A UNIQUE
TRIANGLE"
:Disp "ANGLES AR
E",A,B,C
:Pause
:Disp "SIDES ARE
",D,E,F
:Pause
:Else
:180-B→Q
:B-A→H
:Dsin(H)/sin(A)→
K
:Disp "TWO TRIAN
GLES"
:Disp "PRESS ENT
ER"
:Pause
:Disp "ANGLES ",
A,B,C
:Pause
:Disp "SIDES ",D
,E,F
:Pause
:Disp "ANGLES ",
A,Q,H
:Pause
:Disp "SIDES ",D
,E,K
:End
:End
:Return
:
```

Note: Program SSA uses the If...Then...Else construction to branch among cases.

- There are two End statements at the end of the program. These are needed to complete the two If...Then...Else branchings.
- The pause statements are included after each of the displays to enable you to read the display. To continue, press ENTER .

b) To run SSA, we input the given angle, 40°, the length of the opposite side, 160 m, and the length of the other side, 200 m. We obtain the display shown on the right, indicating that there are two triangles which satisfy these conditions. Upon pressing ENTER twice, we obtain these displays of the angle and side measures of one of the triangles.

```
prgmSSA
GIVEN ANGLE?40
OPPOSITE SIDE?16
0
OTHER SIDE?200
TWO TRIANGLES
PRESS ENTER
```

```
OTHER SIDE?200
TWO TRIANGLES
PRESS ENTER
ANGLES
              40
     53.46414901
     86.53585099
```

```
              40
     53.46414901
     86.53585099
SIDES
             160
             200
     248.4609939
```

Pressing ENTER twice yields the angle and side measures displayed on the right. Depending upon the accuracy of our data, we can round off these measures appropriately.

```
             160
             200
     248.4609939
ANGLES
              40
      126.535851
     13.46414901
```

```
      126.535851
     13.46414901
SIDES
             160
             200
     57.95678338
            Done
```

Triangle 1: Angles: 40°, 53°, 87° Sides: 160 m, 200m, 248 m
Triangle 2: Angles: 40°, 127°, 13° Sides: 160m, 200m, 58 m

WORKED EXAMPLE 3

a) Modify program SAS in *Worked Example* 1, p. 47 to solve a triangle given the measures of two sides and the contained angle.

b) Use SAS to solve the triangle in *Worked Example* 1, p. 47.

Solution

a) A modified version of program, SAS is shown in the display. It prompts for input of the longer side, S, and the other side, T. It then uses the cosine law to compute the third side, R, and the sine law to compute the angle, B. Angle C is calculated using the fact that the sum of the angle measures is 180°.

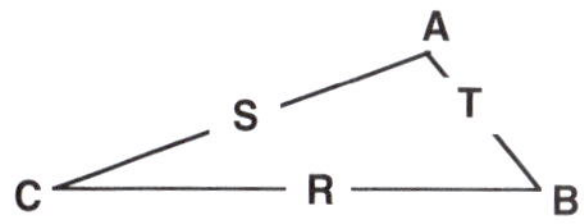

Why did we ask for input of the *longer* side?

The inverse sine, $\sin^{-1}$ returns the acute angle corresponding to an input value of sine. Therefore, we must make sure that we take the inverse sine of an angle which is acute; that is, the angle opposite the smaller side, T.

```
PROGRAM:SAS
:Fix 2
:Degree
:Input "LONGER S
IDE?",S
:Input "OTHER SI
DE?",T
:Input "CONTAINE
D ANGLE?",A
:√(S²+T²-2STcos(
A))→R
:Disp "SIDES ARE
",R,S,T
:Pause
:sin⁻¹(Tsin(A)/R)
→B
:180-A-B→C
:Disp "ANGLES AR
E",A,B,C
:Return
:
```

b) When we run SAS, we obtain the following displays.

```
OTHER SIDE?168
CONTAINED ANGLE?
149
SIDES ARE
          411.21
          258.00
          168.00
```

```
          258.00
          168.00
ANGLES ARE
          149.00
           12.15
           18.85
            Done
```

Exercises

1. Enter program SSA into your calculator. Use it to find the third side of a triangle which has a side of length 2.76 m opposite an angle of 53.38° and another side of length 3.32 m.

2. Can you prove the sine law in a few short lines?

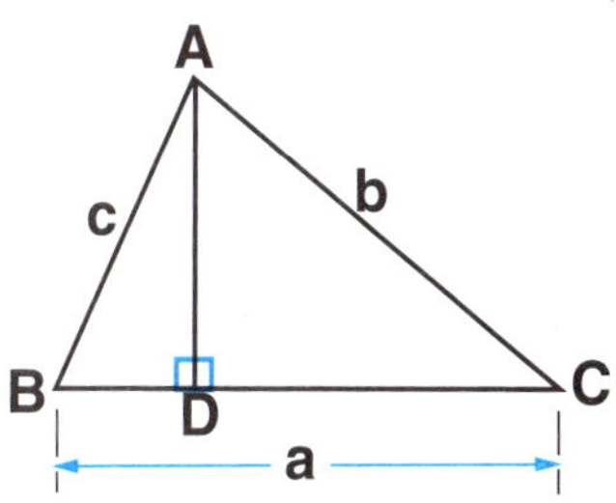

a) Write an expression for the length of AD in ΔABC, in terms of: (i) angle C and length b (ii) angle B and length c.
b) Equate the two expressions in part a)
c) Is your equation true if ΔABC is an obtuse-angled triangle?
d) Use parts a), b) and c) to prove the sine law.

3. Solve ΔABC in exercise 2 if the measures of angles B and C are respectively, 64° and 38° and b = 21.6.

Investigations

4. Communication satellites are located at distances of 128 km and 167 km from an observer on earth. If the angle between the satellites measured by the observer is 27°, what is the distance between the two satellites?

5. To measure the height of a mountain, a crew of surveyors measured its angle of elevation from two positions 300 m apart and in line with its peak. The angles of elevation from the two positions were 37.6° and 49.4°. How high was the mountain peak?

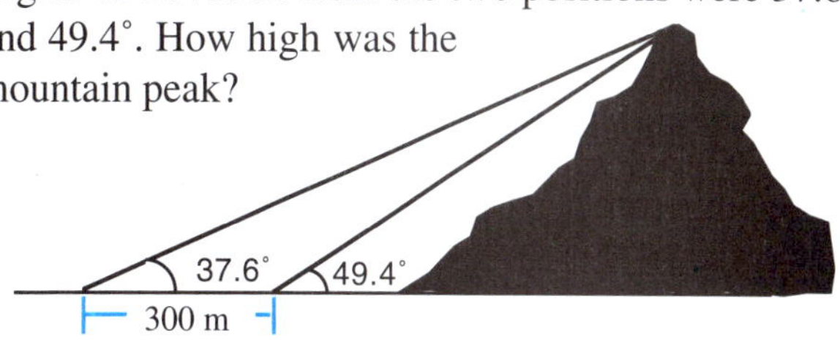

6. Write a program which takes any combination of 3 side lengths or angles of a given triangle as input and yields the measures of all the angles and sides of the triangle. Call the programs SAS, ALLSIDES, AAS, and SSA as subroutines.
Use your program to solve ΔABCwhere:
a) ∠A = 80°; ∠B = 60°; a = 17.6
b) ∠C = 40°; b = 26.5; c = 22.3
c) ∠B = 118°; a =68.7; c = 42.3

Here ∠ means "angle measure."

EXPLORATION 13 The Trigonometric Functions

The motions of the earth and the moon in their axial rotations and orbital revolutions generate the alternation of day and night, the phases of the moon and the cycle of the seasons. Any such change characterized by cycles that repeat over a fixed interval of time is referred to as *periodic*. The time required for a complete cycle is called the *period* of the change. For example, we say that the period of the earth's orbit around the sun is one year.

A Sunrise in New Orleans

Circular or near-circular motion is at the heart of much of the periodicity found in Nature. For this reason the trigonometric (circular) functions are useful in describing many natural phenomena. The nearly circular orbit of the earth combined with the slope of its axis of rotation, enable us to use trigonometric functions to express for any particular location on earth, the number of hours of daylight on any particular day of the year. For example, the number of hours of daylight, y, on the xth day of the year in New Orleans is given by the equation:

$$y=\frac{35}{3}+\frac{7\sin\left(\frac{72(x-80)}{73}\right)}{3}$$

where $x = 1$ on January 1 and $x = 365$ on December 31 and where the argument of the sine function is expressed in degrees.

WORKED EXAMPLE 1

Using the equation above:

a) Graph the number of hours of daylight in New Orleans as a function of the day of the year.

b) Determine the dates on which there is maximum and minimum number of hours of daylight.

c) On approximately what dates will the sun set around 6 o'clock standard time?

Solution

a) To enter the equation above into memory, we use the following keying sequence:

Y = | 3 | 5 | ÷ | 3 | + | (| 7 | ÷ | 3 |) | SIN | 7 | 2 | (| X,T,θ,n | − | 8 | 0 |) | ÷ | 7 | 3 |) | ENTER

WORKED EXAMPLE 1 (cont'd)

Since x is to run from 1 to 365 (all the days of the year) and the number of hours of daylight lies between 5 hours and 15 hours, we press **WINDOW** and set the following limits:

$$1 \le x \le 365 \quad \text{and} \quad 5 \le y \le 15$$

To ensure one period corresponds to $x = 365$, select degree mode by pressing: **MODE** **▼** **▼** **▶** **ENTER**

Then press **GRAPH** to graph the equation. We obtain the display shown.

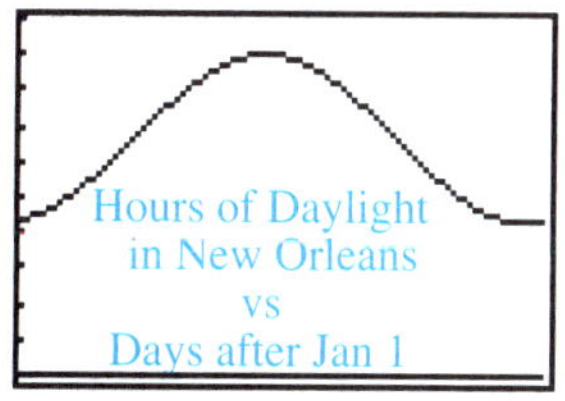

b) To determine the date of maximum daylight, press: **2nd** [CALC] **4** and use the cursor keys to enter values for the lower and upper bounds and the "guess". Our display shows that a local maximum occurs at $x = 171.25$ and $y = 14$. That is, the maximum sunlight is achieved between Day 171 and Day 172. Therefore the day with approximately 14 hours of sunlight is the 172nd day of the year, i.e. June 21.

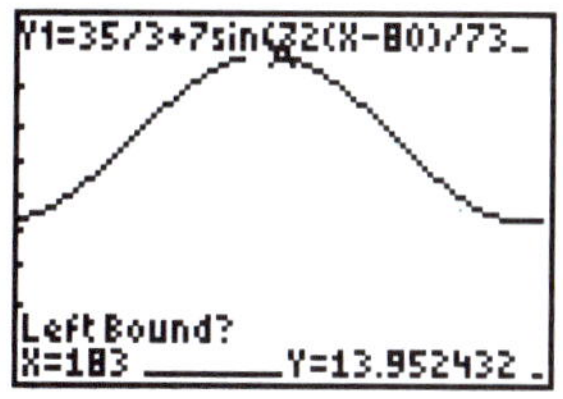

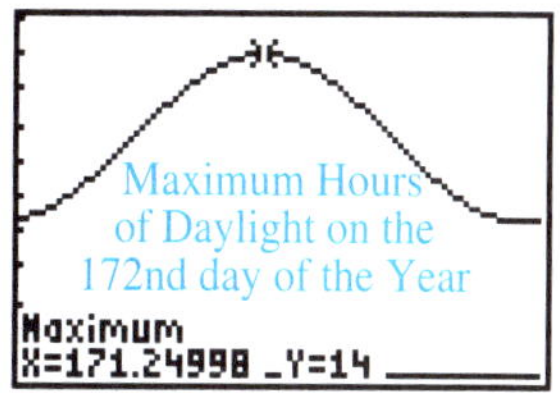

To determine the date of minimum sunlight, we proceed as above after pressing **2nd** [CALC] **3** and enter the lower and upper bounds for the minimum by tracing along the curve toward the minimum so that the graph scrolls horizontally. We obtain the display that indicates a minimum at $x = 353.75$ and $y = 9.33$. This indicates that there are about 9 hours and 20 minutes of daylight on December 20 (the 354th day of the year).

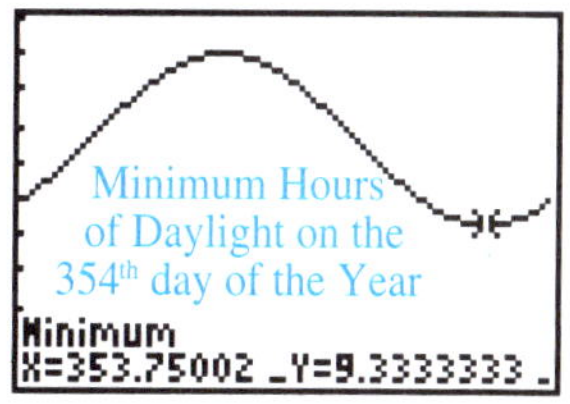

c) The sun sets 6 hours after noon when there are about 12 hours of daylight. By tracing along the graph (press **TRACE**) and zooming in, (press **ZOOM** **2** **ENTER**) we find y is nearest 12 for $x = 88$ and $x = 254$; that is, on March 29 and September 11. (Alternatively, we could find the intersection point with the line $y = 12$.)

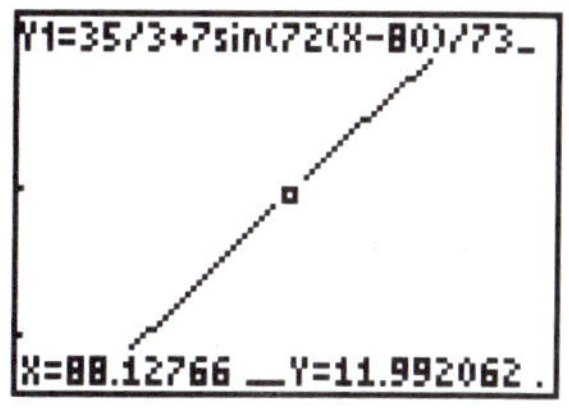

WORKED EXAMPLE 2

The general form of the sine function is: $y = A\sin(Bx + C)$.
Graph the family of sine functions defined by $y = 3\sin Bx$ where $B = 0.5$, 1, 2, and 3.
Then describe how changing B changes the graph.

Solution

To graph this family of curves we define $L_1 = \{0.5, 1, 2, 3\}$, and press:

2nd [{] **.** **5** **,** **1** **,** **2** **,** **3** **2nd** [}] **STO▶** **2nd** [L_1] **ENTER** To enter the functions, press: **Y=** **3** **SIN** **2nd** [L_1] **X,T,θ,n** **ENTER**

To graph in an appropriate viewing window, press: **ZOOM** **7**
We obtain the graphs shown in the display. We observe that when B increases by a factor k, the cycle length is reduced by a factor $\frac{1}{k}$.

Graphs of $y = A\sin(Bx + C)$; $B = 0.5, 1, 2, 3$

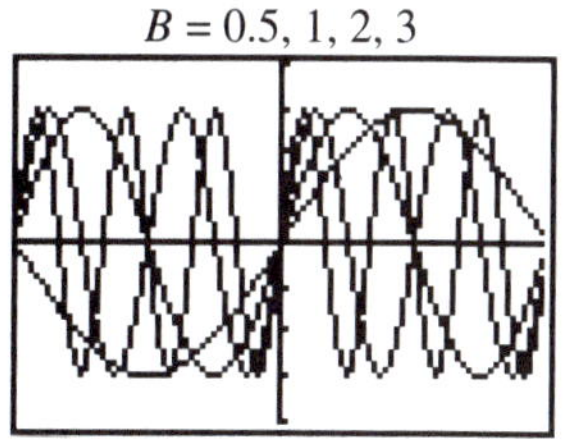

Exercises

1. a) Define a radian. How many radians in a full turn?
b) Write equations for converting degrees to radians and for converting radians to degrees.

2. Write the equation for the number of daylight hours in New Orleans in which the argument of the sine function is given in radians instead of degrees.

For all remaining exercises, convert to radian mode.
Press: MODE ▼ ▼ ENTER

3. Use this keying sequence to graph $y = \sin x$.

Y= SIN X,T,θ,n ENTER GRAPH

Then press ZOOM 7.

This graphs the function with the TrigZoom settings for the window variables.

Press WINDOW to see the new window settings.

4. Using the procedure in *exercise* 3, graph these functions.
a) $y = \sin x$ b) $y = 2\sin x$ c) $y = 3\sin x$
How does the graph of $y = A\sin x$ ($A > 0$) change as A is increased? Decreased? $|A|$ is called the *amplitude*.

5. A *cycle* of a periodic function is a portion of the graph from one point to the point at which the graph starts repeating itself (e.g. the portion of a sine curve between consecutive maxima is a cycle.)
The *period, p,* of a periodic function is the change in x corresponding to one cycle. If p is the period of $f(x)$, then

$$f(x + p) = f(x) \text{ for all } x.$$

Write the period of each function.
a) $y = \sin 4x$ b) $y = \cos 2x$ c) $y = 8\cos 0.2x$

6. Graph the following functions using the ZoomTrig window settings.
a) $y = \sin x$ b) $y = \sin\left(x+\frac{\pi}{3}\right)$ c) $y = \sin\left(x+\frac{\pi}{2}\right)$

Note: To enter the value π, use 2nd [π].

Describe what happens to the graph of the function $y = \sin(x + C)$ as C changes. $-2\pi \le C \le 2\pi$

Investigations

7. Graph the functions $y = \sin\left(x+\frac{\pi}{2}\right)$ and $y = \cos x$ using the ZoomTrig window. Display both graphs simultaneously. Write an equation which relates the sine function and the cosine function. Graph the functions defined by $y = \sin x$ and $y = \cos\left(x-\frac{\pi}{2}\right)$ Write an equation relating the sine and cosine functions. Are both equations the same? Explain.

8. The diagram shows the graphs of the functions defined by $f(x) = \sin x$ and $g(x) = \sin\left(x-\frac{\pi}{3}\right)$. The graph of $g(x)$ is the same as the graph of $f(x)$ except it is shifted right by $\frac{\pi}{3}$ radians. This horizontal displacement of the sine curve is called the *phase displacement* or *phase shift*. We say that $g(x)$ has a phase shift of $\frac{\pi}{3}$.

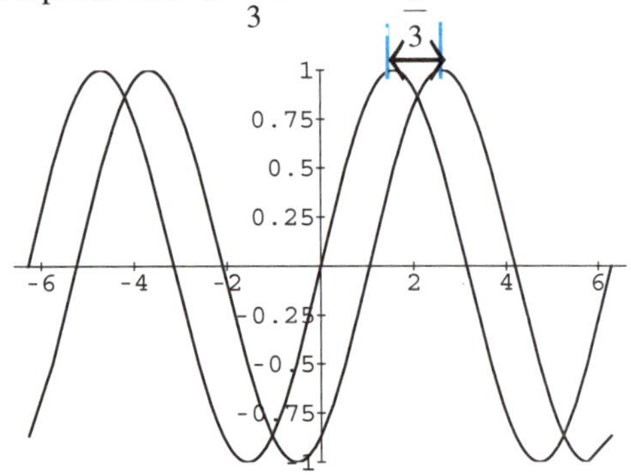

Use the results of your previous investigation to predict the amplitude, period, and phase shift of these functions. Sketch graphs to check your answers.

a) $\sin\left(x-\frac{\pi}{6}\right)$ b) $2\sin\left(\frac{x}{2}+\pi\right)$ c) $-3\sin\left(2x+\frac{\pi}{3}\right)$

HOW MUCH DAYLIGHT WHERE YOU LIVE?

Use an almanac to determine the times of sunrise and sunset on June 21 (longest day of the year) and December 21 (shortest day) in the city closest to where you live.

Assume that the number of hours of daylight on the xth day of the year is given by an equation of the form:

$$y = A\sin(Bx + C) + D$$

Calculate the values of the constants, A, B, C, and D when the argument of the sine function is taken to be in degrees.
Write the equation when the argument of the sine function is taken to be in radians.

Graph the equation and use the trace function to predict the times of sunrise and sunset today. Check these answers with the times posted in the newspaper. Why might these times differ?

EXPLORATION 14 The Exponential Function

In April 1968, a group of 30 scientists assembled in Rome to identify the major problems threatening the survival of the human race. This group, which later became the *Club of Rome*, identified the world population explosion as the foremost danger facing modern civilization. The world population had reached 3 billion in 1959; it subsequently reached 4 billion in 1974; 5 billion in 1986, and 6 billion in 1997. It has taken from the beginning of time until 1959 to reach 3 billion and only 38 years to increase another 3 billion!

When the *rate* of growth of a particular variable increases in proportion to the current value of the variable, we say the growth is *exponential*. The graph on page 58 shows the world population during the past two centuries. The rate of population increase (represented by the slope of the curve) becomes greater (steeper) as the population grows.

To provide a mathematical model for exponential growth, we define an *exponential function* of x to be a function described by an equation of the form:

$$y = b^{Ax+C}$$

where A and C are constants and b is a positive constant. The example below shows how the exponential function has been used to model world population growth. The weakness in this model is that it predicts unlimited human population over time, so the model is valid for only restricted periods. The cartoon illustrates the two major models for exponential change — growth and decay!

© 1996 by Sidney Harris

When the constant A is negative, the exponential function defined by $y = b^{Ax+C}$ describes what is called *exponential decay*. In this case, the *rate* at which the function decreases is proportional to its current value. For example, when a radioactive substance decays, its mass decreases as it emits radioactive particles. The rate of decay is proportional to the amount of material present. It can be shown (see *worked example* 2) that in such cases, the time required for 50% of the substance to decay is fixed. This time is called the *half life*.

WORKED EXAMPLE 1

The world population y is expressed as a function of the year x by the equation $y = 10^{0.00389x+2}$.

a) Graph this equation. Trace along your graph to estimate the years in which the population reached 1 billion, 3 billion, and 5 billion people.

b) Predict the year when the population will reach 7 billion.

SOLUTION

a) We enter: Y= 1 0 ^ (**0.00389** X,T,θ,n + 2) ENTER

Then we set the window values as shown in the display and press GRAPH.

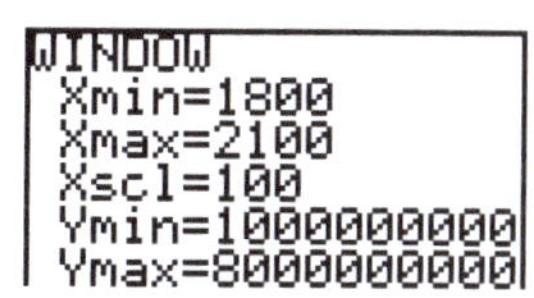

We obtain the display shown below. Using the trace function, we obtain the years, 1800, 1922,1979 and 1999 as the approximate years when the world population reached 1, 3, 5 and 6 billion. If a more precise estimate is needed, we can proceed as in part b).

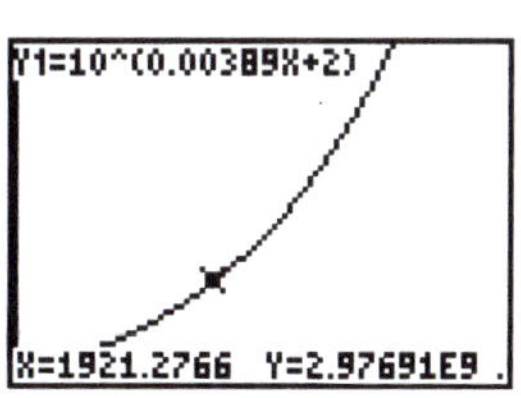

b) To estimate when the population will reach 7 billion, we define $Y_2 = 7000000000$ and enter

2nd [CALC] 5

We find the graphs intersect before $x = 2017$. That is, the population is predicted to reach 7 billion in 2016.

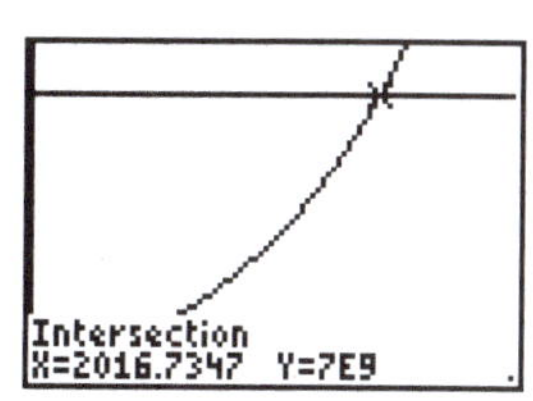

WORKED EXAMPLE 2

Carbon 14 is an isotope of carbon found in plants that have absorbed it from the atmosphere. From the time the carbon 14 is absorbed, it decays exponentially. The percentage, y, of carbon 14 in the plant x years after absorption is given by the following equation.

$$y = 10^{2-0.00005235x}$$

Graph this equation and determine the half-life of carbon 14.

Solution

To graph this exponential function, we enter:

[Y=] [1] [0] [^] [(] [2] [−] **0.00005235** [X,T,θ,n] [)] [ENTER]

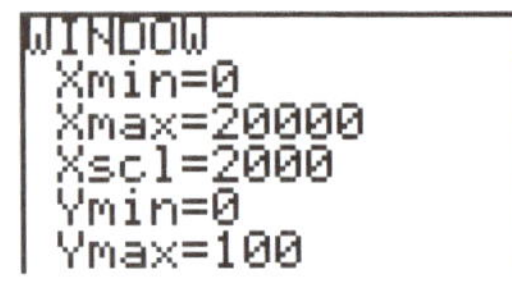

Then we set the range values as shown in the display and press [GRAPH] to obtain the graph shown in the display.

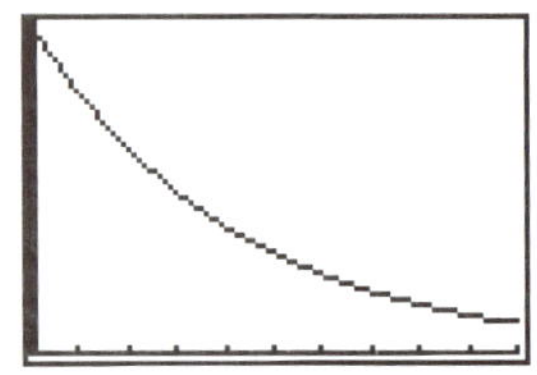

To determine the half-life of carbon 14, we must find the value of x for which $y = 50$ since that is the number of years it takes for the carbon 14 isotope to reach 50% of its initial mass. To obtain a precise value of the half-life, we can graph the line $y = 50$ and solve for the intersection with the graph shown in the display. To find this point of intersection, we enter

[2nd] [CALC] [5]

and after responding to the prompts, we obtain the display showing that the half-life of carbon 14 is about 5750 years.

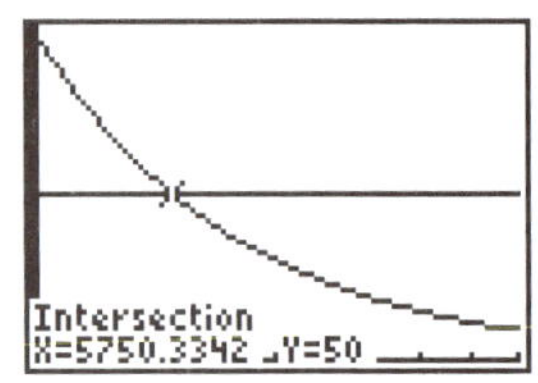

An alternate method

Alternatively, we could have traced along the curve by pressing [TRACE]. Pressing the [◀] key repeatedly, we move the cursor along the curve to the point with coordinates (5744.6809, 50.034084). This indicates that the half-life is about 5745 years. To obtain a closer approximation, we could zoom in further or define a new rectangular window using the **ZBox** option on the [ZOOM] menu as described below.

To Use the ZBox Option

- Press [ZOOM] [ENTER]
- Move the cursor two spaces up and two spaces left using the cursor keys. Then press [ENTER]. This defines the upper left corner of the rectangular window you are about to draw.
- Move the cursor 5 spaces right and 5 spaces down and press [ENTER].
- The graph is redrawn in the new window you have drawn.
- Press [TRACE] and move the cursor along the curve to (5749.2078, 50.006789).
- From this it follows that the half life is close to 5749 years.

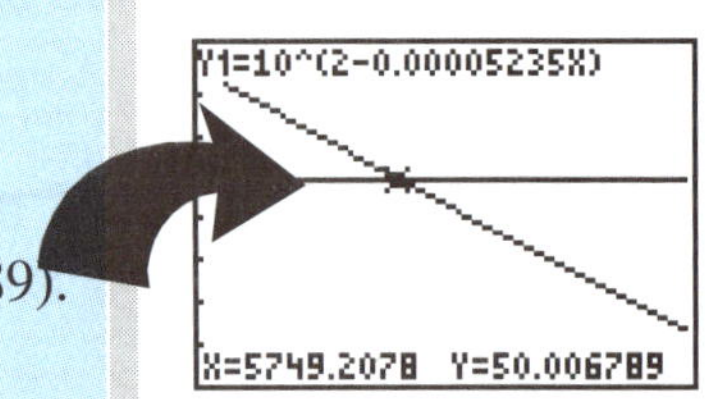

If we use the ZBox option one more time, we can trace along the curve to the point, (5750.0506, 50.00171) indicating that, to the closest integer, the half-life is 5750 years.

Exercises

1. Describe what happens to the slope of the graph of the function $f(x) = b^{Ax}$ (where $A > 0$) as:

a) $x \to \infty$ b) $x \to -\infty$

2. Describe what happens to the slope of the graph of the function $f(x) = b^{Ax}$ (where $A < 0$) as:

a) $x \to \infty$ b) $x \to -\infty$

3. A radioactive substance decays exponentially with a half-life of 500 years. How long will it take the entire substance to decay? Explain your answer.

4. Graph the function $f(x) = 8^{Ax}$ for this set of values of A. $A = \{-1, -0.5, -0.25, 0.25, 0.5, 1\}$

a) Describe how the graph of $f(x)$ changes as A changes and as $|A|$ changes.

b) Show algebraically how the y-coordinate of a point on the graph of $f(x) = 8^x$ is related to the point vertically above or below it, on the graph of $f(x) = 8^{0.5x}$.

c) Write the equation of the image of the graph of $f(x) = 8^{Ax}$ under a reflection in the y-axis.

5. Given the function defined by $f(x) = b^{Ax}$ where $b > 1$, $A > 0$ and x takes non-negative real values:

a) Prove that there is a *doubling time d* such that:

$f(x + d) = f(x)$ for all x.

b) Write an equation relating A, b, and d which does not involve x.

c) The function defined by $f(x) = (1.16)^x$, gives the accumulated value at the end of x years of \$1 invested at 16% per year and compounded annually. Calculate the doubling time of $f(x)$.

6. Given the function defined by $f(x) = b^{Ax}$ where $b > 1$, $A < 0$ and t takes non-negative real values:

a) Prove that there is a *half-life h* such that:

$$f(t + h) = \frac{1}{2} f(t) \quad \text{for all } t$$

b) Write an equation relating A, b and h which does not involve t.

c) The proportion of an isotope of radium which is undisintegrated after x years of radioactive decay is given by the function $M(x)$ where $M(x) = e^{-0.0004332x}$ ($e \approx 2.718$). Graph this function for $0 < x < 4000$ and $0 < y < 1$ and determine the half-life of radium.

Investigations

7. In *worked example* 1, we represented the world population, y, in the year x (A. D.) by the equation:

$$y = 10^{0.00389x + 2}$$

Graph this equation and find the doubling time.

The table below shows the approximate years in which the world population reached each billion according to the U.S. Census Bureau. (Visit www.census.gov/ipc/www/worldpop.html.)

Year	1801	1925	1959	1974	1986	1998
Population in Billions	1	2	3	4	5	6

How long did it take the world population to double from 1 billion to 2 billion? Is this more or less than the doubling time of the exponential function above?

How long did it take the population to double from 2 to 4 billion? How close is this to the doubling time computed above? How long did it take to double from 3 to 6 billion? Compare this with the doubling time predicted by the exponential equation? Is the world population growing exponentially? Is its growth exceeding exponential growth? Explain. When will the world population reach 8 billion?

8. Another mathematical model for population growth uses the equation

$y = e^{0.0173x - 12}$ where $e = 2.718...$

Graph this equation. Determine its doubling period. Does this give a better approximation to the 1997 prediction than the equation in part a)? For what year do the two equations give the same prediction?

CHALLENGE

Given $f(x) = a^{4x} - 5a^{2x} - 6$ where $a > 0$

a) Express $f(x)$ as a product.

b) Does the equation $f(x) = 0$ have a root?

c) Find all roots of the equation $f(x) = 0$ when $a = 3$.

EXPLORATION 15 The Logarithmic Function

WHEN WILL THE WORLD'S POPULATION BE DOUBLE ITS PRESENT SIZE?

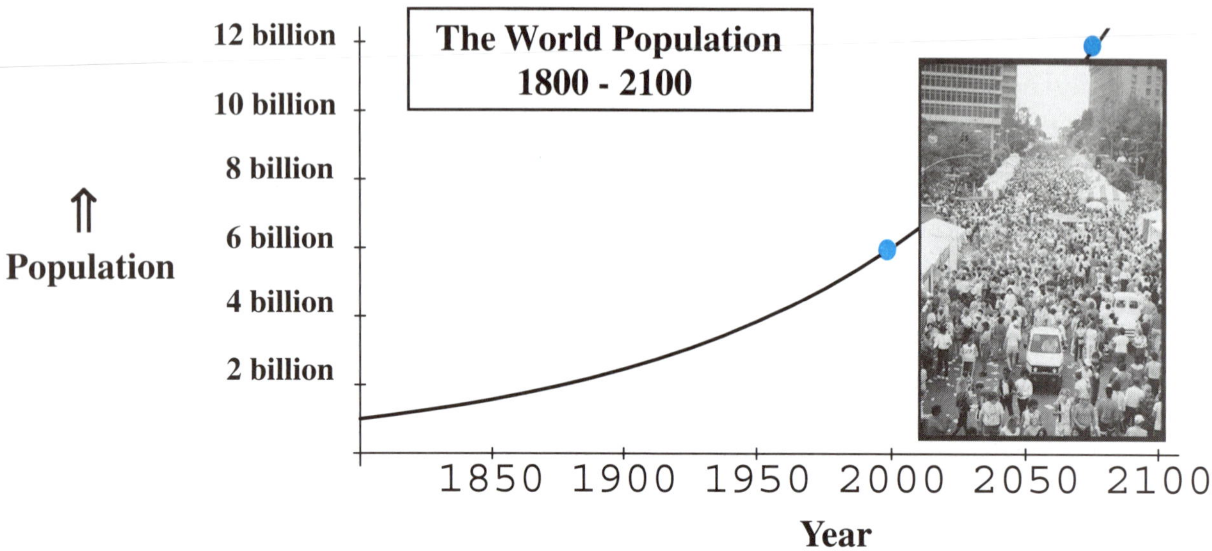

How long will it take the world population to double from 6 billion to 12 billion? In *Exploration* 14, we expressed the world population y as a function of the year x where the function is defined by the equation:

$$y = 10^{0.00389x + 2}$$

To find the year x corresponding to $y = 12$ billion, we can find the point of intersection of this graph with the graph of $y = 12$. Alternatively, we can solve the following equation for x:

$$12 \times 10^{9} = 10^{0.00389x + 2}$$

Multiplication of both sides by 10^{-9} yields:

$$12 = 10^{0.00389x - 7}$$

To solve this equation, we must express 12 as a power of 10 so that we can equate the exponents on both sides of the equation. The exponent to which 10 must be raised to yield 12 is called *the logarithum of* 12.

We observe: $10^2 = 100$; we write: $\log 100 = 2$
We observe: $10^3 = 1000$; we write: $\log 1000 = 3$

> In general, we define the (common) logarithmic function of x, which we denote "$\log x$", by the equation:
>
> $$10^{\log x} = x$$

To find the (common) logarithm of a number using the TI-84, we press the LOG key followed by the number.

The sequence LOG 1 2 ENTER yields 1.07918... That is, $\log 12 = 1.07918\ldots$
We can therefore substitute $10^{1.079}$ for 12 in the equation above to obtain:

$$10^{1.079} \approx 10^{0.00389x - 7}$$

Equating exponents and solving for x, yields $x \approx 2077$. Therefore, if the world population is accurately described by the equation $y = 10^{0.00389x+2}$, then it will reach 12 billion around the year 2077 — doubling in less than a century! Could the model be wrong?

"Yes, yes, I *know* that, Sidney—*everybody* knows *that*! ... But look: Four wrongs *squared*, minus two wrongs to the fourth power, divided by this formula, *do* make a right."

WORKED EXAMPLE

Graph a) The function $f(x) = \log x$ where $0 < x \leq 100$.
b) The year as a function of the world population for the period between 1800 and the year 2100.
c) In about what year will the population will reach 10 billion?.

Solution

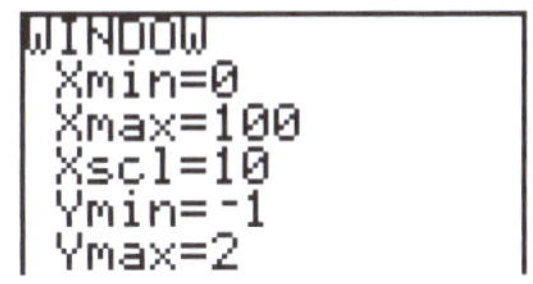

a) To graph $\log x$ as a function of x, we first set the range. Since $\log x$ is defined only for $x > 0$, and since $\log 100 = 2$, we set the window values as in the display.

With these window settings in place, we press these keys:

Y= **LOG** **X,T,θ,n** **ENTER** **GRAPH**

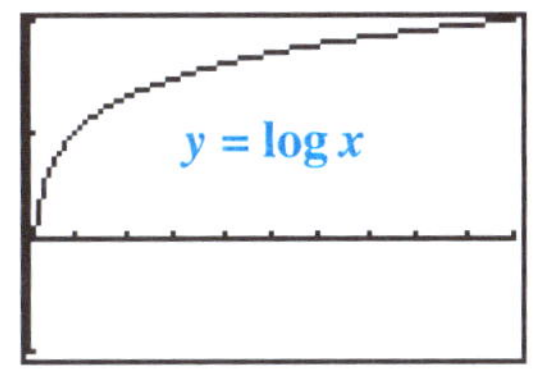

and we obtain this (monotonically) increasing graph showing that $\log x$ increases as x increases.

b) To plot the year as a function of the world population, we solve the equation

$$y = 10^{0.00389x + 2}$$

for x by taking the logarithm of both sides of the equation and solving the linear equation in x to obtain

$$x = \frac{\log y - 2}{0.00389} \quad \text{or} \quad x \approx 257(\log y - 2)$$

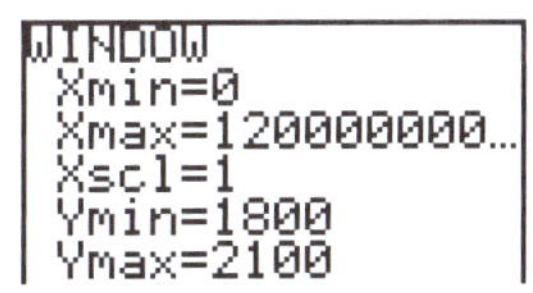

Since the TI-84 Plus calculator graphs y as a function of x, we replace y with x to graph the expression above and set the range values as shown in the display.

Y= **2** **5** **7** **(** **LOG** **X,T,θ,n** **)** **−** **2** **ENTER** **GRAPH**

This yields the graph shown here.

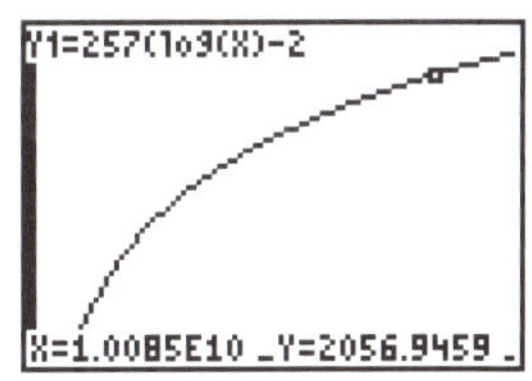

c) We press **TRACE** and use the **▶** key to trace along the curve to the point,

X= 1.0085E10, Y= 2056.9459

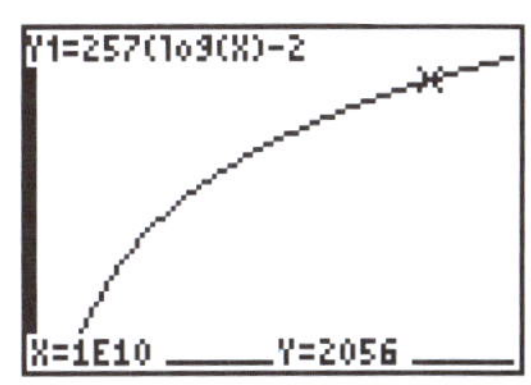

Alternatively, we press: **2nd** [CALC] **ENTER** and in response to **X=** we enter 10000000000. This produces the display indicating **Y = 2056**.

That is, a population of 10 billion is reached around 2056.

Note: The prediction in part "c" is based on the exponential model for population growth given by the equation, $y = 10^{0.00389x+2}$. The prediction is only as valid as this model and it implies that the world population will continue to grow without limit. Such an exponential model for population growth was originally advanced by Thomas Malthus in 1798. This model had a dramatic influence on the social philosophy of the time. It was argued by some, that war, disease and starvation were necessary evils to prevent overpopulation. Some of these ideas influenced Charles Darwin in his formulation of the concept of "survival of the fittest". In 1845, the Belgian scientist P. F. Verhulst formulated a model for population growth that does not increase without limit. (See Exploration 16.)

Exercises

1. Write a number for each of the following without using your calculator. Then use your calculator to check
a) log 1000 b) log 100000 c) log 0.001

2. a) Write the equation $y = \log x$ in exponential form. (That is, express x as a power of 10.)
b) Determine the value of x if $\log x = 4$.

3. Use your calculator to evaluate each expression.
a) log 2 b) log 20 c) log 200 d) log 2000

4. Study your answers in *exercise* 3. Explain how the logarithm of 2 is related to the logarithm of a number which is 10, 100 or 1000 times as large?

5. a) If $a = \log 2$ and $b = \log 20$, use a and b to write 2 and 20 as powers of 10. Use an exponent law to express b in terms of a.

b) Compare your result in part (a) with what you discovered in exercise 4.

c) Conjecture how the graphs of the functions $\log x$ and $\log 10x$ might be related?

6. a) Graph each of the following functions in the range:
$0 < x \le 100$; $-1 \le y \le 5$
i) $\log x$ ii) $\log 10x$ iii) $\log 100x$

b) Use the TRACE key to move between the graphs you have plotted. Observe the change in the y-coordinates as you move vertically. Explain what you discover.

7. Solve each equation without graphing.
a) $\log x = 2.4$ b) $\log x = 3.8$
c) $\log x = 0.8$ d) $\log x = -2.4$

8. For any non-zero value of x, evaluate $\log x + \log x^{-1}$. Explain what you discover.

9. a) Graph the functions $Y_1 = \log x$ and the function $Y_2 = \log 5x$ in the range: $0 < x \le 10$; $-1 \le y \le 2$

b) Press these keys to define $Y_3 = Y_1 - Y_2$ and graph it.

Y= VARS ▶ ENTER 1 −
VARS ▶ ENTER 2 ENTER GRAPH

c) Trace along the graph of Y_3. Describe what all the points on this graph have in common.

d) Conjecture how the graphs of $y = \log x$ and $y = \log Ax$ (where A is a positive constant) are related.

Investigations

10. a) Graph the functions $Y_1 = \log x + \log 5$ and $Y_2 = \log 5x$ in the range: $0 < x \le 10$; $-1 \le y \le 2$.

b) Press these keys to define and graph $Y_3 = Y_1/Y_2$.

Y= VARS ▶ ENTER 1 ÷
VARS ▶ ENTER 2 ENTER GRAPH

c) Trace along the graph of Y_3. Describe what all the points on this graph have in common.

d) Conjecture a relationship between the graphs of $y = \log x + \log A$ and $y = \log Ax$ where A is any positive constant.

11. Using the definition of a logarithm, and the exponent laws, prove that for any positive real numbers $\boldsymbol{a}$ and $\boldsymbol{b}$:
(i) $\log \boldsymbol{ab} = \log \boldsymbol{a} + \log \boldsymbol{b}$
(ii) $\log \boldsymbol{a/b} = \log \boldsymbol{a} - \log \boldsymbol{b}$

12. a) Use your TI-84 Plus to find the value of log 4.

b) Use your value of log 4 to deduce these logarithms.
(i) log 40 (ii) log 400 (iii) log 4000
(iv) $\log \frac{1}{4}$ (v) $\log \frac{1}{40}$ (vi) $\log \frac{1}{400}$
(vii) log 2.5 (viii) log 25 (ix) log 250

13. a) Graph these functions:
$Y_1 = 2 \log x$, $Y_2 = \log x^2$ and $Y_3 = Y_1/Y_2$
in the range: $0 < x \le 10$; $-1 \le y \le 2$.

b) Use the trace function to explore the graph of Y_3.

c) Explain what you discovered about the graphs of Y_1 and Y_2.

14. a) Use the method in *exercise* 13 to compare the graphs of $Y_1 = 3 \log x$ and $Y_2 = \log x^3$.

b) Use your discoveries in part (a) to conjecture a relationship between the graphs of $Y_1 = n \log x$ and $Y_2 = \log x^n$, where n is a positive integer.

c) Test your conjecture graphically for various n.

d) For what values of n other than positive integral values is your conjecture true?

15. a) Determine the value of log 2 to three decimal places.

b) Use the relationship you discovered in *exercise* 14, and the value of log 2 to compute these logarithms.
(i) log 4 (ii) log 8 (iii) log 16
(iv) log 64 (v) log 6400 (vi) log 16000

Exercises

16. Solve the following equations without using your calculator.

a) $\log x + \log x^4 = 15$ b) $\log \sqrt[4]{x} = 1.25$

c) $2\log \sqrt[3]{x} = -2$ d) $\log \sqrt[3]{x} = \frac{2}{3}$

17. Solve each of the following equations for x by taking the logarithm of both sides and isolating x.

a) $5^x = 100$ b) $4^{x-3} = 10$ c) $3^{2x-1} = 100$

18. Describe how the graph of $f(x)$ defined by: $f(x) = A\log\left(\frac{x}{B}\right)$, changes as A increases; as B increases.

19. Find a root of the equation $\log x = \frac{2}{3}$ by graphing the logarithmic function and tracing along the curve. Check your answer by computing $10^{2/3}$.

20. Solve the equation $3 \log x = 7 \log (x - 2)$ for x, to two decimal places, by graphing the equations $y = 3 \log x$ and $y = 7 \log (x - 2)$ and finding the point of intersection.

21. a) Graph the functions $Y_1 = \log x$ and $Y_2 = 10^x$. in the range: $-5 < x \le 10$; $-3.4 \le y \le 3.4$.
b) If • denotes function composition, describe the graphs of the functions $Y_1 \bullet Y_2$ and $Y_2 \bullet Y_1$.
c) If two functions $f(x)$ and $g(x)$ and satisfy the identity: $f(x) \bullet g(x) = x$ for all x, then $f(x)$ and $g(x)$ are called *inverse* functions. What is the inverse of $\log x$?

22. Solve for x, $10^{2x} - 2(10^x) + 1 = 4$.

23. Determine the minumum value of 8^{x^2+2x}.

24. The points $A(x_1, y_1)$ and $B(x_2, y_2)$ are two points on the graph of $y = \log x$. Through the midpoint of the line segment AB, a horizontal line is drawn to cut the curve at $C(x_3,y_3)$. Prove that $x_3^2 = x_1x_2$.

CHALLENGE

x, y, and z are three real numbers such that:

$$0 < x < 1, \quad y = x^x, \quad \text{and} \quad z = x^y$$

Write the three numbers x, y, and z in order of increasing magnitude.

Investigations

25. The function $f(x) = \ln x$ (pronounced "lawn of x") is the *natural logarithm* of x; that is, $\ln x$ is defined to be the inverse of the function e^x where $e = 2.71828\ldots$

a) Use the **LN** key on your TI-84 Plus to evaluate:
i) ln 5 ii) ln 50 iii) ln 1 iv) ln 10

b) By definition, $\ln x$ satisfies the equations, $\ln(e^x) = x$ and $e^{\ln x} = x$. Use either or both of these equations to prove these properties of the natural logarithm for positive numbers a and b.
(i) $\ln ab = \ln a + \ln b$ (ii) $\ln a/b = \ln a - \ln b$
(iii) $\ln a^b = b \ln a$ (iv) $\ln \sqrt[b]{a} = \frac{\ln a}{b}$

c) Without using your TI-84, solve for x: $e^x - e^{-x} = 1$.
d) Use your TI-84 Plus to check your answer in part c).

26. a) Graph $y = \log x$ and $y = \ln x$. Using your graphs, conjecture the value of the constant k such that $\ln x = k \log x$
b) Using the definitions of the natural and the common logarithms, express the value of the constant k in terms of e.

HOW GOOD IS THE PRIME NUMBER THEOREM?

In 1896, two French mathematicians, Jacques Hadamard and Charles-Jean de la Vallée Poussin, proved the following celebrated *Prime Number Theorem*

> If $\pi(n)$ denotes the number of prime numbers between 0 and n, then as n gets large,
> $$\pi(n) \rightarrow \frac{n}{\ln n}$$

Use this theorem to estimate the number of primes up to:
a) 100 b) 1000 c) 10,000
d) 100,000 e) 1,000,000

The actual number of primes is respectively, 25, 168, 1229, 9592 and 78,498. List the percentage error beside each estimate. Does the percentage error decrease as $n \rightarrow \infty$?

Graph $\pi(x)$ for $0 \le x \le 10{,}000{,}000$. Use your graph to estimate $\pi(10{,}000{,}000)$. Search the Internet or the library to find the actual value of $\pi(10{,}000{,}000)$. What is the percentage error in your estimate?

EXPLORATION 16 Limits & Asymptotic Behavior

In Explorations 14 and 15, we used an exponential function to model the world human population as a function of time. In this model, as x increases without limit (we write: $x \to \infty$) the world population $f(x)$ increases without limit (we write: $f(x) \to \infty$). However, we know from our study of populations of insects, birds and mammals that once the population of a species reaches the limits of its food supply, it approaches a stable equilibrium. We say that the population $f(x)$ approaches a *limiting value* L as x approaches infinity. We write:

$$\lim_{x \to \infty} f(x) = L$$

In the 1845, P. F. Verhurst modeled world-population growth using the *logistic function*, defined by an equation of the form:

$$f(x) = \frac{A}{1 + ae^{-cAx}}$$

where A ,a and c and are positive constants and e is a special positive mathematical constant which we explore in the investigations.

In the special case, when $A = 1$, $a = 1$ and $c = 1$, the logistic function is defined by:

$$y = \frac{1}{1 + e^{-x}}$$

The graph of this function is shown in the display for these window settings: $-5 \le x \le 8$; $-1 \le y \le 2$.

To find the limiting value of the population, in the general case, we observe that as x increases without limit, the exponential function, ae^{-cAx} approaches closer and closer to zero.
(We say: ae^{-cAx} approaches zero *asymptotically*.) Therefore the denominator of $f(x)$ approaches 1 asymptotically and so $f(x)$ approaches the limiting value A asymptotically.

We write: $\lim_{x \to \infty} f(x) = A$

That is, as time increases the population described by the logistic function above, approaches the limiting value A asymptotically. The logistic function has been found more useful than the exponential function in describing not only population growth but other dynamic systems such as chemical kinetics and ecology where asymptotic behavior is important.

Larson's cartoon gives a somewhat different meaning to the term *asymptotic behavior.*

More recently, the Verhulst model of population growth has been described by the *logistic difference equation,*

$$x_{n+1} = \lambda x_n (1 - x_n)$$

This equation has been studied recently in the new branch of mathematics popularly called *Chaos Theory*. For a more detailed investigation of this equation using the TI-84 Plus, see *Advanced Algebra with the TI-84 Plus Calculator,* published by *Brendan Kelly Publishing Inc.*

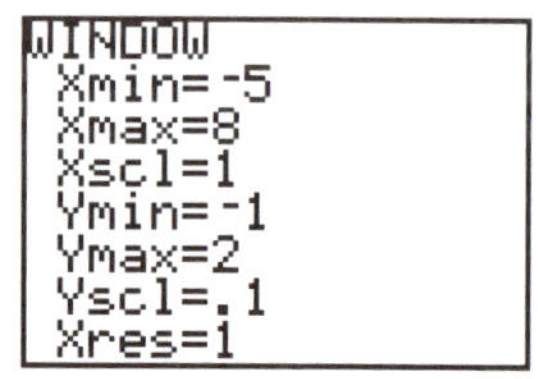

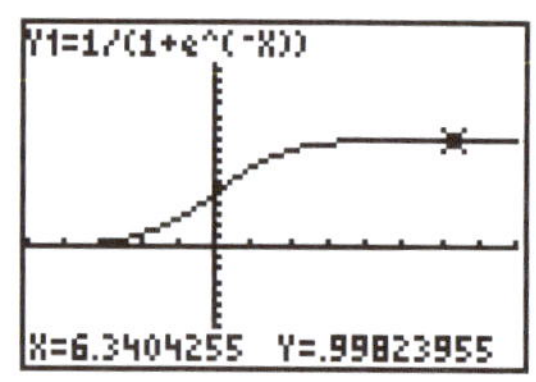

THE FAR SIDE® By GARY LARSON

WORKED EXAMPLE

A sum of \$10 000 invested at 12% per annum grows to

$$\$10\,000\left(1+\frac{0.12}{k}\right)^{20k}$$

over a period of 20 years if interest is compounded k times per year. What is the accumulated value of \$10 000 after 20 years if money is compounded:

a) annually? b) semi-annually? c) monthly?
d) daily? e) instantaneously?

Solution

We evaluate the formula at these values of k:

a) $k = 1$, to obtain $\$10\,000(1.12)^{20} \approx \$96\,462.93$.
b) $k = 2$, to obtain $\$10\,000(1.06)^{40} \approx \$102\,857.18$.
c) $k = 12$, to obtain $\$10\,000(1.01)^{240} \approx \$108\,925.54$.
d) $k = 365$, to obtain $\$10\,000(1.000328\ldots)^{7300} \approx \$110\,188.29$.
e) To compound *instantaneously* means to compound at the rate that is equivalent to compounding over an infinitesimally short period. To determine the result of instantaneous compounding, we must compute:

$$\lim_{k\to\infty} \$10\,000\left(1+\frac{0.12}{k}\right)^{20k}$$

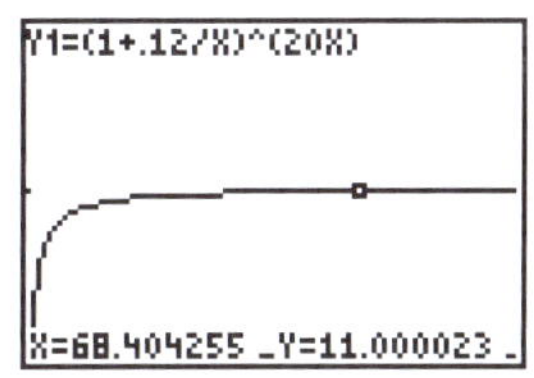

To investigate this limit, we graph the function $f(x)=\left(1+\frac{0.12}{x}\right)^{20x}$ in the range $1 \le x \le 100$; $10 \le y \le 12$.
We see in the display that as x increases the graph appears to be a horizontal line. Tracing to the right on successive redrawings of the graph becomes tedious. No matter how far we trace, we will never get the limit in the window. In such a case it is useful to substitute $\frac{1}{x}$ for x and evaluate the limit as $x \to 0$.

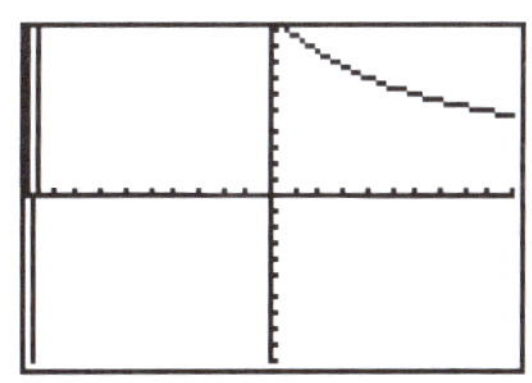

That is, we graph the function $f(x)$ where

$$f(x)=(1+0.12x)^{\frac{20}{x}}$$

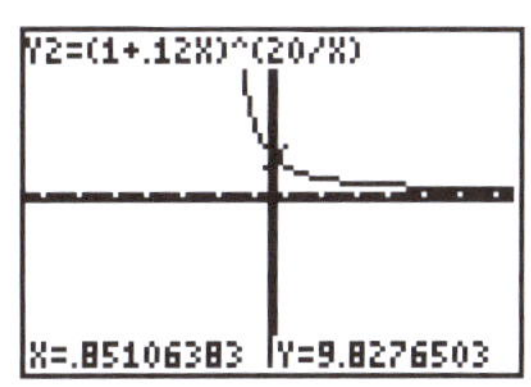

With the default range settings, we obtain a graph in the first quadrant of the screen. Upon zooming out, we obtain the graph shown opposite. We trace along this graph toward the y-axis until we reach in turn, each of the points:

(0.85106383, 9.8276503) and (-0.8510638, 12.573525).

These bounding points prompt us to reset the window to $-0.0001 \le x \le 0.0001$ and $9.8 \le y \le 12.5$. Pressing **GRAPH** **TRACE** ▶

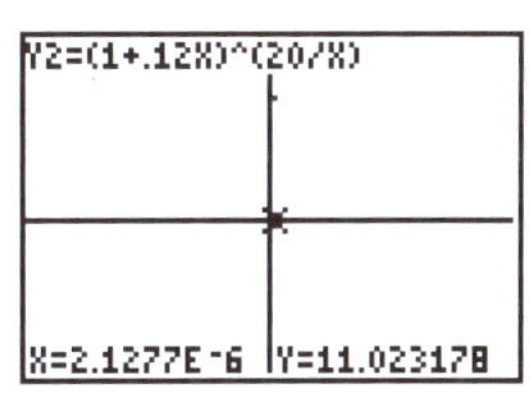

places the cursor at the point, (2.1277 E-6, 11.023178) so,

$$\lim_{k\to\infty}\left(1+\frac{0.12}{k}\right)^{20k} \approx 11.023$$

Therefore, the accumulated value of \$10 000 over 20 years of instantaneous compounding is about \$110 230. Observe that instantaneous compounding does not yield much more than daily compounding over the 20-year period!

Exercises

1. Explain what is meant by *asymptotic behavior.*

2. Describe the asymptotic behavior of each of the graphs below, stating in each case the limit approached by the function as $x \to -\infty$ and as $x \to \infty$.

a) $f(x)=\dfrac{14x^4-3x^2+1}{3x^4+3x+17}$ b) $F(x)=\dfrac{12x^3}{x^4+1}$

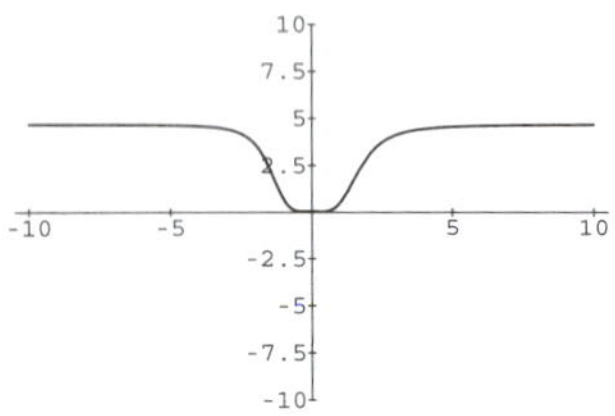

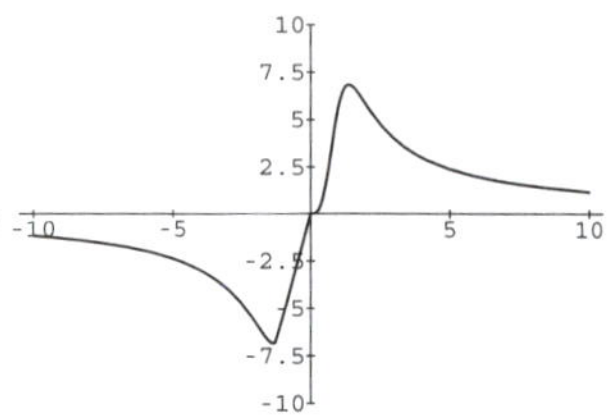

c) $g(x)=\dfrac{17.6}{2+10^{-0.4x}}$ d) $G(x)=7x\sin x^{-1}$

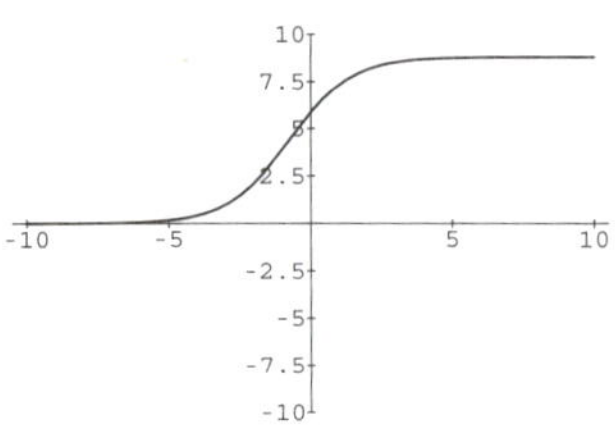

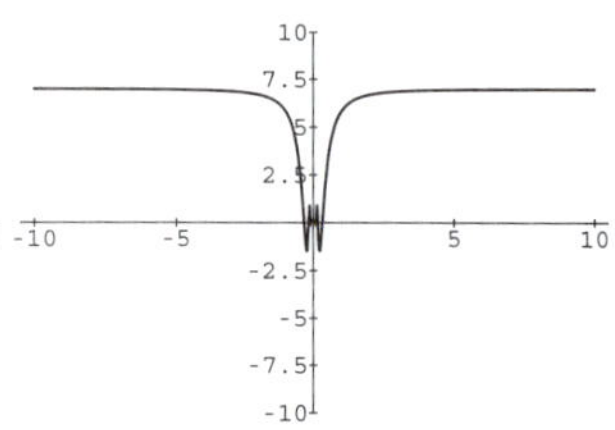

3. Describe the behavior of each of the following functions as $x \to -\infty$ and as $x \to \infty$.

a) $f(x)=3x^2-2$ b) $g(x)=\dfrac{1}{x-3}$

c) $h(x)=10^{-3x}$ d) $m(x)=\sin x$

4. Evaluate each limit (if it exists).

a) $\lim\limits_{x\to\infty} 3x^2-2$ b) $\lim\limits_{x\to\infty}\dfrac{1}{x-3}$

c) $\lim\limits_{x\to\infty} 10^{-3x}$ d) $\lim\limits_{x\to\infty}\sin x$

5. Write expressions for $f\left(\frac{1}{x}\right)$, $g\left(\frac{1}{x}\right)$ and $h\left(\frac{1}{x}\right)$ where $f(x)$, $g(x)$ and $h(x)$ are defined as in *exercise* 3.

6. Use the expressions you found in *exercise* 5 to evaluate:

a) $\lim\limits_{x\to 0} f\left(\frac{1}{x}\right)$ b) $\lim\limits_{x\to 0} g\left(\frac{1}{x}\right)$ c) $\lim\limits_{x\to 0} h\left(\frac{1}{x}\right)$

Exercises

7. Compare your answers in *exercises* 4 and 6 and state a relationship between the asymptotic behavior of a function of x and the behavior of the function as x^{-1} approaches 0.

8. Evaluate using the relationship you discovered in *exercise* 7.

a) $\lim\limits_{x\to\infty}\dfrac{3x^3-4x+2}{2x^3+4x^2-7}$ b) $\lim\limits_{x\to\infty}\dfrac{4x^3-3x^2+1}{2x^5-3x}$

c) $\lim\limits_{x\to\infty} x^{-\frac{3}{2}}$ d) $\lim\limits_{x\to\infty}\left(\dfrac{3x+1}{x}\right)^2$

9. a) Graph the function $f(x)=\dfrac{\sin x}{x}$.

Press **ZOOM** **7** to set the window to the trig function setting.

b) Trace along your graph to evaluate $\lim\limits_{x\to 0} f(x)$.

10. Use your answer in *exercise* 9 to evaluate these limits.

a) $\lim\limits_{x\to 0}\dfrac{\sin 2x}{x}$ b) $\lim\limits_{x\to 0}\dfrac{\sin x}{2x}$

c) $\lim\limits_{x\to 0}\dfrac{\tan 3x}{x}$ d) $\lim\limits_{x\to 0}\dfrac{5x}{\sin 2x}$

11. Evaluate: $\lim\limits_{x\to\infty}\dfrac{(6x+1)^4-(6x)^4}{x^3}$

12. If $F(x)=x^2[1-\cos(x^{-1})]$ evaluate $\lim\limits_{x\to\infty} F(x)$.

13. Evaluate: $\lim\limits_{x\to 0}\dfrac{\sqrt{30x+30}-\sqrt{30}}{x}$.

"UNDERSTANDING CREATION AND INFINITY IS WITHIN OUR GRASP... JUST AS SOON AS EVOLUTION PROVIDES US WITH LARGER BRAINS."

Investigations

14. a) Express $(1+\frac{1}{x})^{cx}$ in terms of $(1+\frac{1}{x})^{x}$.

b) Let **e** denote the limit: $\lim\limits_{x\to\infty}(1+\frac{1}{x})^{x}$

Write $\lim\limits_{x\to\infty}(1+\frac{1}{x})^{cx}$ in terms of *e*.

c) Write $\lim\limits_{x\to\infty}(1+\frac{A}{x})^{x}$ in terms of *e*.

d) Write $\lim\limits_{x\to\infty}(1+\frac{A}{x})^{cx}$ in terms of *e*.

15. a) Graph the function $f(x)=(1+\frac{1}{x})^{x}$.

b) Trace along the graph to evaluate $\lim\limits_{x\to\infty} f(x)$ to 3 decimal places.

c) The limit you have just evaluated is denoted *e* and is a widely used mathematical constant. The second function key above the **LN** key is the [e^x] key.

To check your answer in part b, and obtain the decimal expansion of *e*, press:

2nd [e^x] **1** **ENTER** (This yields e^1 or *e*.)

16. Use your answers from exercise 14 and the [e^x] function on your calculator to calculate these limits.

a) $\lim\limits_{x\to\infty}(1+\frac{1}{x})^{2x}$ b) $\lim\limits_{x\to\infty}(1+\frac{0.14}{x})^{x}$

c) $\lim\limits_{x\to\infty}(1+\frac{0.5}{x})^{4x}$ d) $\lim\limits_{x\to\infty}(1+\frac{1}{x})^{0.5x}$

17. Graph the functions of *x* given in Investigation 16. Use these graphs to investigate the asymptotic behavior of those functions and verify the values of the limits which you obtained in *exercise* 16.

18. Express **e** as the limit of a function as $x\to 0$. Graph that function and verify that **e** is the limit of that function as $x\to 0$.

Investigations

19. One model of world population growth proposes that the world population in *billions* $f(t)$ expressed as a function of the year t, $(t>0)$, is given by:

$$f(t)=\frac{12}{1+e^{-0.032(t-2000)}}$$

a) Use this formula to predict the world population in these years.
(i) 2000 (ii) 2050 (iii) 2080

b) Graph $f(t)$ and trace along the curve to verify your predictions in part a).

c) Trace further along the curve to predict the world population in the year 3000.

d) Does the population approach a limit or will it eventually grow indefinitely large? Explain.

e) If you believe it will approach a limit, evaluate $\lim\limits_{t\to\infty} f(t)$ by tracing along the curve.

f) Evaluate $\lim\limits_{t\to 0} f\left(\frac{1}{t}\right)$ by tracing along the curve defined by $f\left(\frac{1}{t}\right)$. What does this limit tell you?

20. What is the accumulated value of $10 000 invested at 10% over a period of 25 years if it is compounded:
a) annually? b) daily? c) instantaneously?

In his famous paper of 1905, Albert Einstein presented his *Special Theory of Relativity*, which asserts that a clock traveling at a high velocity will be slower than a clock traveling at a lower velocity. This "elongation" of time intervals is called *time dilation.* Einstein's theory implies that if it takes t_0 seconds for a meson to decay when it is at rest, then it will take t seconds for it to decay when traveling at x% of the speed of light, where t is given by:

$$t=\frac{t_0}{\sqrt{1-\left(\frac{x}{100}\right)^2}}$$

A brilliant young woman named Ms. Bright
Could travel much faster than light
She departed one day,
In an Einsteinian way,
And came back on the previous night.

a) Graph t/t_0 as a function of x for: $0\le x<100$; $0\le y\le 10$

b) By what factor is the decay time dilated for a meson traveling at
(i) 90% the speed of light? (ii) at 99.90% the speed of light?

c) Determine the speeds at which the decay time for a meson is:
(i) doubled (ii) tripled (iii) increased 50-fold

d) Describe the decay time as a meson approaches the speed of light.

THE BETTMANN ARCHIVE

Galileo Galilei 1564 – 1642

© 1997 by Sidney Harris

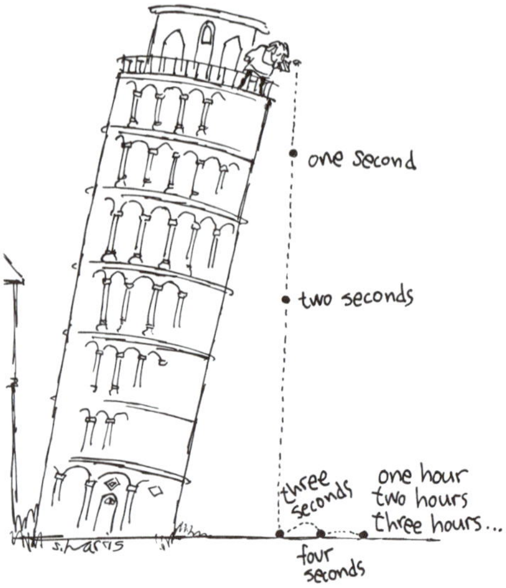

When Isaac Newton made his famous statement, *If I have seen further than others, it is because I have stood on the shoulders of giants*, he was probably thinking of both Archimedes and Galileo among others. Galileo Galilei, who died in the very year that Newton was born, is considered to be the father of modern experimental physics. In his study of moving bodies, he stated informally the principles later embodied in Newton's first two laws of motion. Galileo was the first person to use the telescope to study the heavens. His observations proved (in conflict with religious doctrine) that the solar system is heliocentric. For this he was convicted of heresy by the Inquisition in Rome, ordered to recant and placed under house arrest for the remainder of his life. (The Pope recently issued an apology to Galileo.) His work paved the way for Newton and others to develop the laws of motion and formulate the mathematics of orbits and trajectories.

Galileo had asserted that all objects accelerate to earth at a uniform rate and this rate is the same for all objects no matter what their mass.

The following table shows the total distance covered after t seconds (for $0 \le t \le 9$) by an object falling in a vaccuum.

Distance in Meters traveled by a Falling Object in *t* Seconds										
Time (s)	0	1	2	3	4	5	6	7	8	9
Distance (m)	0	4.9	19.6	44.1	78.4	122.5	176.4	240.1	313.6	396.9

If Galileo were correct, then the average acceleration over each one-second interval should be the same for all intervals. To determine the average acceleration, we create a table showing the average velocity over each one-second interval by subtracting consecutive distances in the table above. We obtain the following table.

Average Velocity of a Falling Object in the tth Second										
Time (s)	0	1	2	3	4	5	6	7	8	9
Av. Velocity (m/s)	0	4.9	14.7	24.5	34.3	44.1	53.9	63.7	73.5	83.3

Subtraction of average velocities for consecutive intervals yields the following table.

Average Acceleration of a Falling Object in the tth Second										
Time (s)	0	1	2	3	4	5	6	7	8	9
Av. Acceleration (m/s^2)	9.8	9.8	9.8	9.8	9.8	9.8	9.8	9.8	9.8	9.8

We observe that the object accelerated at a uniform rate of 9.8 m/s^2. Galileo was correct.

WORKED EXAMPLE 1

a) Using the fact that a free falling object has uniform acceleration of 9.8 m/s^2, express the distance s traveled by the object as a function of the time of falling t.

b) Graph the distance vs. time curve and find the distance traveled in the interval $2.8 \le t \le 9.2$. Find the average velocity of the object during this time interval using the distance- time graph.

SOLUTION

a) Acceleration is the rate of change of velocity. Since the acceleration is fixed at 9.8 m/s^2, the velocity is increasing at the constant rate of 9.8 m/s^2. Therefore the velocity after t seconds is $9.8t$ meters / second. This is shown in the diagram.

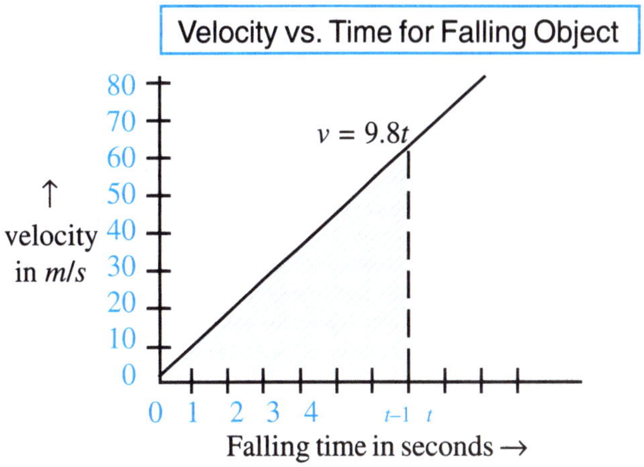

The distance traveled in the first t seconds of free fall is represented by the area under the velocity vs. time curve.

The distance traveled by an object during any time interval is its average velocity (in *meters per second*) during that time interval, times the number of *seconds* in the time interval. Therefore the distance traveled by the object in the first t seconds of free fall is:

$$\underbrace{\frac{9.8t - 0}{2}}_{\text{average velocity}} \times \underbrace{t}_{\text{time}} \quad \text{or} \quad 4.9t^2$$

We observe that the total distance $s(t)$ is given by $s(t) = 4.9t^2$. This is also represented by the area under the velocity-time graph in the interval $[0, t]$. Galileo used this idea in 1638.

Distance vs. Time

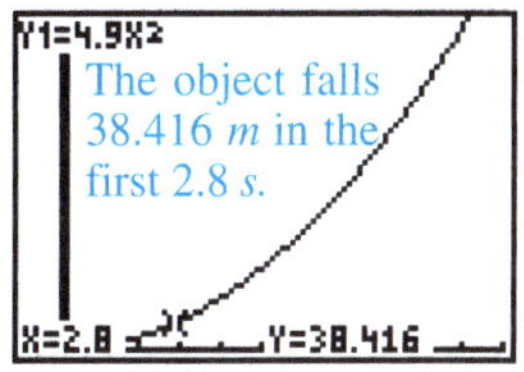

b) To graph this function, we define $Y_1(x) = 4.9x^2$ Then we graph this function, in the window: $-1 \le x \le 11; -1 \le y \le 500$. To find the value of the distance traveled when $t = 2.8$ s, (i.e. $x = 2.8$), we select **value** from the CALC menu, by pressing:

2nd [CALC] **ENTER**

In response to the prompt, **X =** , we press **2.8** **ENTER** .
The display gives the corresponding distance, 38.416 m.

Distance vs. Time

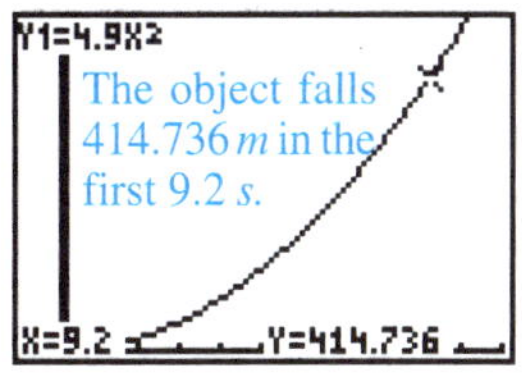

To repeat this procedure for $t = 9.2$, press: **2nd** [CALC] **ENTER** .
In response to the prompt **X =** we enter **9.2** **ENTER** .
The display yields the corresponding distance, 414.736 m.
The change in distance fallen between $t = 2.8$ s and $t = 9.2$ s is Δs, where $\Delta s = 414.736 - 38.416 = 376.32$ m.

$$\text{Average velocity} = \frac{\Delta s}{\Delta t} \begin{array}{l}\leftarrow \text{change in distance} \\ \leftarrow \text{change in time}\end{array}$$

$$= \frac{376.32}{6.4} \quad \text{or } 58.8 \; m/s$$

Distance vs. Time

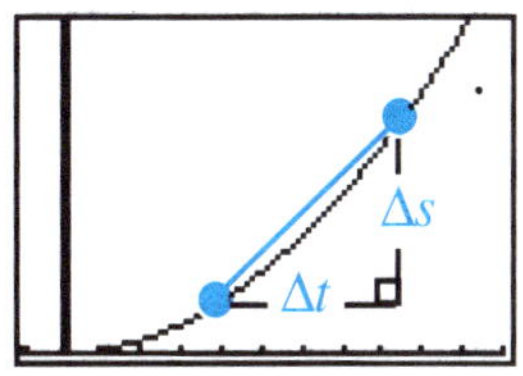

The slope of the secant gives the average velocity over the interval Δt.

We observe that the average velocity over any time interval is just the slope of the corresponding secant on the distance-time graph.

In *worked example* 1, we saw that the average velocity over a time interval from t_0 to $t_0 + \Delta t$, is given by the slope $\Delta s/\Delta t$ of the secant joining $(t_0, s(t_0))$ to $(t_0 + \Delta t, s(t_0 + \Delta t))$. As we shorten the length Δt of the time interval, (that is, as $\Delta t \to 0$) the slope of the secant becomes the slope of the tangent at $(t_0, s(t_0))$.

As $\Delta t \to 0$ the average velocity over the time interval from t_0 to $t_0 + \Delta t$ becomes the *instantaneous velocity* at time t_0. This instantaneous velocity is given by the slope of the *tangent* at the point $(t_0, s(t_0))$. The slope of this tangent is denoted by the symbol:

$$\left.\frac{ds}{dt}\right|_{t_0}$$

This symbol is read, *the derivative of s(t) with respect to t at the point with t = t_0*. In geometric terms, $\left.\frac{ds}{dt}\right|_{t_0}$ denotes the slope of the curve at the point $(t_0, s(t_0))$. The formal definition of the derivative as the limit of the slope of the secant as $\Delta t \to 0$, is:

$$\lim_{\Delta t \to 0} \frac{\Delta s}{\Delta t} = \frac{ds}{dt}$$

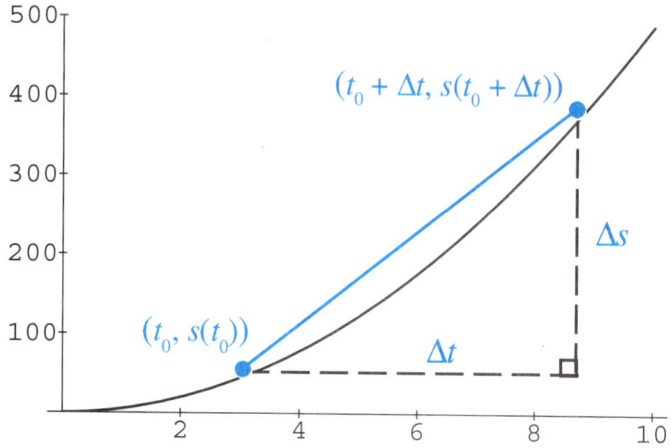

The following *Worked Example* shows how we can evaluate the instantaneous velocity at t_0 from a time-distance curve by calculating $\left.\frac{ds}{dt}\right|_{t_0}$ on the TI-84 Plus.

WORKED EXAMPLE 2

On August 1, 1996 at the Summer Olympics in Atlanta, Georgia, American sprinter Michael Johnson ran the 200-m dash in 19.32 seconds – shattering the previous Olympic and World Records by a significant margin. That record has survived two Olympic Games and remains a world record as this book goes to press. The distance *s* meters, traveled by Johnson in the first *t* seconds of this race is closely approximated by the cubic polynomial:

$$s(t) = -0.0419t^3 + 1.239t^2 + 2.048t + 0.443$$

Calculate Johnson's velocity at:

(i) $t = 7$ (ii) $t = 12$ (iii) $t = 18$.

SOLUTION

We press **Y=** and enter the function:

Y1=-.0419X³ + 1.239X² + 2.048X + .443

Then we set the window variables to :

$0 \le x \le 20;\ 0 \le y \le 200$

and press: **GRAPH** to obtain this display.

To obtain Johnson's velocity at $t = 7$, press:

2nd [CALC] **6** **7** **ENTER**

The display shows that the slope of the distance-time curve at $t = 7$ is about 13.2, so Johnson's velocity at the 7-second mark was about 13.2 *m/s*.

Similarly, we find his velocities at $t = 12s$ and 18 *s* are respectively 13.7 *m/s* and 5.9 *m/s*.

AGENCE FRANCE PRESSE/ CORBIS-BETTMANN

Exercises

1. a) State a mathematical relationship relating Δs, Δt and v_{av} given that Δs is the change in distance traveled by a projectile during a time interval Δt, and v_{av} is its average velocity during that time interval.

b) Describe how you can determine v_{av} for any time interval using a distance-time graph.

2. a) State a mathematical relationship relating Δv, Δt and a_{av} given that Δv is the change in velocity of a projectile during a time interval Δt, and a_{av} is its average acceleration during that time interval.

b) Describe how you can determine a_{av} for any time interval using a velocity-time graph.

c) Using the velocity-time graph in *worked example* 1, state the average velocity over the time interval, $2.8 \leq t \leq 9.2$.

3. If an object moves so that the acceleration is constant, does this mean that the velocity is a linear function of time? Explain why or why not.

4. a) Suppose a cannon ball were dropped from the top of the leaning tower of Pisa (a height of 56 *m*). How long would it take for it to reach the ground, if we ignore the effect of wind resistance?

b) The cannon ball would have fallen 44.1 *m* after 3 seconds and 78.4 *m* after 4 seconds. Would the distance fallen after 3.5 seconds be the average of these two distances? Explain why or why not.

5. Graph the distance-time equation given in *worked example* 2 for Michael Johnson in the time interval $0 \leq t \leq 20$.

a) Use the **Value** command on the CALC menu to determine the distances traveled by Johnson at $t = 10$ seconds and $t = 15$ seconds.

b) Calculate Michael Johnson's average velocity in the interval $10 \leq t \leq 15$.

c) Use the method in *worked example* 2 to calculate Johnson's instantaneous velocity at $t = 10$ seconds.

d) Consider your answer in part b) to be an estimate of Johnson's velocity at $t = 10$ seconds. Compare this estimate of the instantaneous velocity at $t = 10$ seconds with the actual velocity which you found in part c).

Investigations

6. The vertical distance s traveled by an object in free fall for t seconds, is given by $s = 4.9t^2$.

a) Write an expression for $s + \Delta s$, the distance traveled by the object in the first $t + \Delta t$ seconds.

b) Use your answer in part a) to help you write (as a function of t) an expression for Δs, the distance fallen in the interval from t to $t + \Delta t$.

c) Write an expression in terms of t, and Δt for the average velocity during the time interval Δt.

d) Evaluate $\lim_{\Delta t \to 0} \frac{\Delta s}{\Delta t}$.

e) Use your answer to part d) to estimate the velocity of an object in free fall, 4.5 seconds after it has been released.

7. The distance y (in meters) traveled by a Dodge VIPER R/T accelerating from a standing start during a power test, is expressed in terms of the time t (in seconds) by the equation:

$$y = -0.07t^3 + 3.15t^2 + 1.2t$$

VIPER photo courtesy of Steve Smith

a) Graph the distance-time curve for the VIPER during the first 10 seconds of the road test.

b) Determine the total distance traveled in the first t seconds, for:

(i) $t = 3$ seconds (ii) $t = 6$ seconds
(iii) $t = 8$ seconds (iv) $t = 10$ seconds

c) Calculate the average speed of the VIPER between $t = 3$ and $t = 6$.

d) Calculate the instantaneous speed of the VIPER at $t = 3$ seconds.

e) At what time in the first 10 seconds, did the VIPER reach its maximum speed?

EXPLORATION 18 *From f(x)unctions to Relations – The Ellipse*

Prior to the 17th century, people did not understand how the earth moved in relation to the sun. Though Copernicus had asserted in 1543 that the earth was not stationary, his theory was not generally accepted until scientists like Galileo and Tycho Brahe gathered more evidence. Furthermore, the shape of the earth's orbit and the orbits of the other planets around the sun was still unknown until Johannes Kepler enunciated his three laws of planetary motion in publications dated 1609 and 1619. Kepler's laws were empirical and it remained for scientists to explain in terms of some theory why these laws were true.

Johannes Kepler 1571-1630

© 1997 by Sidney Harris

Kepler's Three Laws

1. *Each planet moves around the sun in an elliptical orbit. The sun is located at a focus of the ellipse.*
2. *Each planet travels around the sun in such a way that the line joining the planet to the sun sweeps through equal areas in equal times.*
3. *If P_1 and P_2 are the orbital periods of two planets, and R_1 and R_2 the respective mean distances from the sun, then:*

$$\left(\frac{P_1}{P_2}\right)^2 = \left(\frac{R_1}{R_2}\right)^3$$

WHAT IS AN ELLIPSE?

Any curve defined by an equation of the form,

$$\frac{x^2}{a^2} + \frac{y^2}{b^2} = 1$$

where a and b are both constants, is an *ellipse*.

When $a > b$:

the major (longer) axis of the ellipse lies along the x-axis.

a is the semi-major axis

b is the semi-minor axis

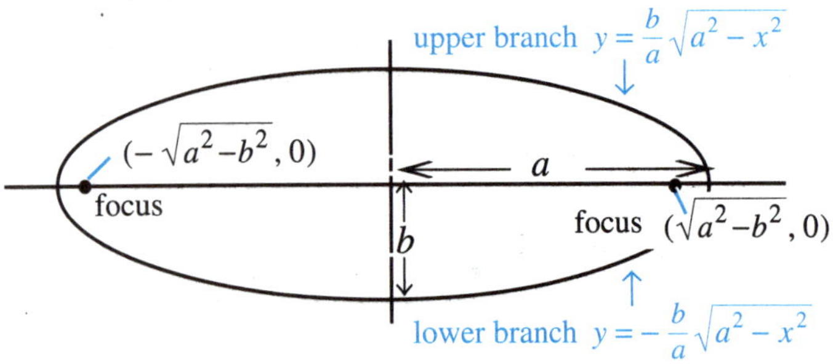

When $a < b$:

the major (longer) axis of the ellipse lies along the y-axis.

b is the semi-major axis

a is the semi-minor axis

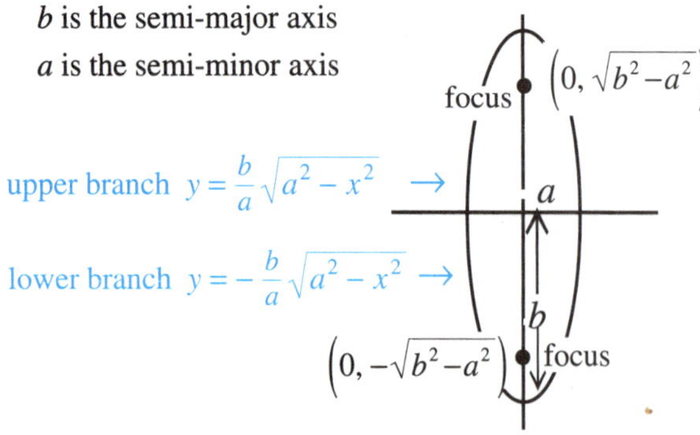

A NEW VISION OF OUR SOLAR SYSTEM

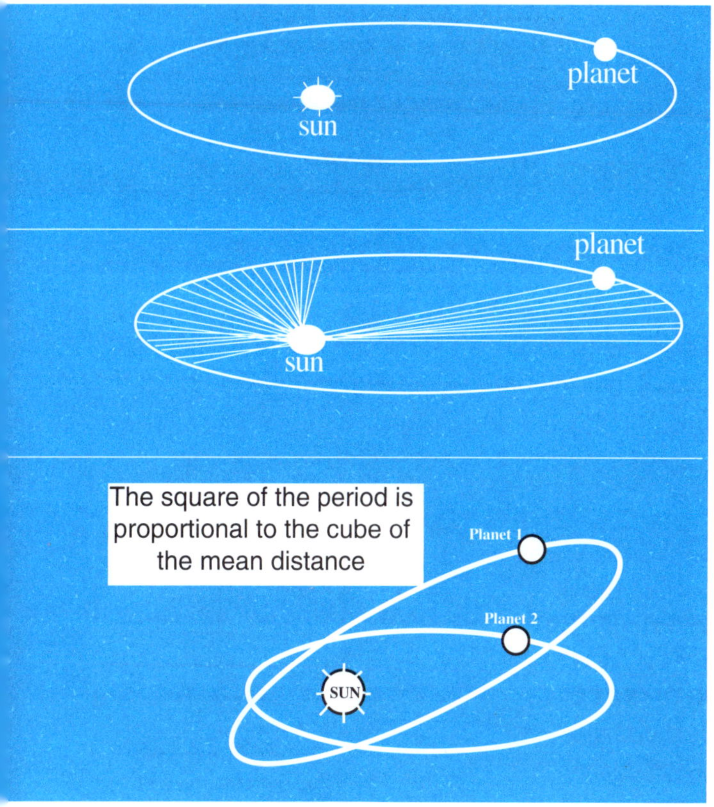

In 1663-64, Cambridge University was closed on account of the bubonic plague. Isaac Newton, who was a twenty-one year old undergraduate, was sent home to continue his studies. During that highly productive period, Newton formulated his law of universal gravitation which states that the force of attraction between any two masses is proportional to the product of their masses and inversely proportional to the square of the distance between them. During this time, Newton also developed a new branch of mathematics called *calculus*. Using the methods of calculus, he was able to deduce the three laws of Kepler from his own law of universal gravitation. He waited until 1686 to publish this work in a treatise titled *Philosophiae Naturalis Principia Mathematica*. The scientist, Pierre-Simon Laplace, exclaimed,

Isaac Newton 1642-1727

Newton was surely the man of genius par excellence, but we must agree that he was also the luckiest: one finds only once the system of the world to be solved!

WORKED EXAMPLE 1

Derive the equation of the curve traced out by a point moving so that the sum of its distances from the points $F_1(-c, 0)$ and $F_2(c, 0)$ is a constant k. Identify the curve from its equation.

SOLUTION

The condition that the sum of the distances of a point $P(x, y)$ from F_1 and F_2 is k units can be written $PF_1 + PF_2 = k$,

i.e. $$\sqrt{(x+c)^2+y^2}+\sqrt{(x-c)^2+y^2}=k \quad ①$$

Arranging the radicals so they appear on opposite sides of equation ① yields:

$$\sqrt{(x+c)^2+y^2}=k-\sqrt{(x-c)^2+y^2} \quad ②$$

Squaring both sides of ② yields: $(x+c)^2+y^2=(x-c)^2+y^2-2k\sqrt{(x-c)^2+y^2}+k^2$ ③

Collecting terms and simplifying ③, we obtain: $k^2-4cx=2k\sqrt{(x-c)^2+y^2}$ ④

Squaring both sides of ④ and collecting like terms yields: $\dfrac{x^2}{\left(\frac{k}{2}\right)^2}+\dfrac{y^2}{\left(\frac{k}{2}\right)^2-c^2}=1$ ⑤

Equation ⑤ is of the form $\dfrac{x^2}{a^2}+\dfrac{y^2}{b^2}=1$ with $a = k/2$ and $b=\sqrt{a^2-c^2}$. Therefore the path traced out by a point moving so that its total distance from $F_1(-c, 0)$ and $F_2(c, 0)$ is a constant k, is an ellipse with $a = k/2$ and $b=\sqrt{a^2-c^2}$. Points F_1 and F_2 are called *foci* (plural of *focus*) of the ellipse.

WORKED EXAMPLE 2

Graph the ellipses defined by the equations:

$16x^2 + 49y^2 = 784$ ①

and $64x^2 + 9y^2 = 576$ ②

Write the coordinates of the points of intersection of the two ellipses.

> **NOTE:** A VERTICAL LINE THROUGH THE CENTER OF AN ELLIPSE INTERSECTS THE ELLIPSE IN TWO POINTS. THIS INDICATES THAT AN ELLIPSE IS NOT THE GRAPH OF A FUNCTION. WHY?

SOLUTION

Dividing equation ① by the constant term yields: $\frac{x^2}{7^2} + \frac{y^2}{4^2} = 1$.

Comparing this equation with standard form, we see that $a = 7$ and $b = 4$.
That is, this equation defines an ellipse with semi-major axis 7 and semi-minor axis 4.
Since $a > b$, the major axis lies along the x-axis.

Similarly, we divide ② by the constant term, 576 to obtain: $\frac{x^2}{3^2} + \frac{y^2}{8^2} = 1$,
from which we deduce, $a = 3$ and $b = 8$.
Since $a < b$, the major axis lies along the y-axis.

We can use the values of a and b to sketch both ellipses and then solve algebraically for their intersection points. (See exercise 6 p.73.) However, when the values of a and b are not simple integers, it is often more convenient to graph the ellipses and use the **intersect** command on the CALC menu as shown on page 23. This yields the intersection point (2.6598…, 3.6999…) in the first quadrant. The coordinates of the other three points are deduced by symmetry.

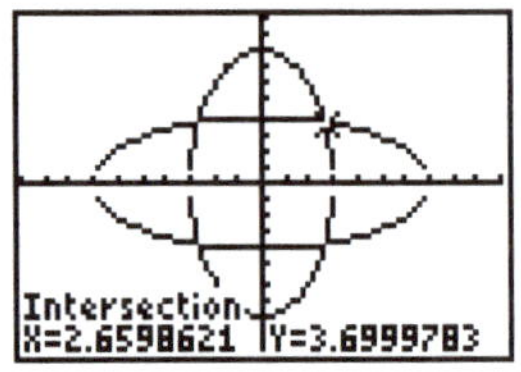

Ellipses displayed with the default window settings.

WORKED EXAMPLE 3

Neptune has an elliptical orbit with major axis 60.14 A. U. and minor axis 60.12 A.U.

Graph the orbit of Neptune relative to the sun at the origin.

> An A. U. (astronomical unit) is the mean distance of the earth from the sun and is equal to approximately 149.7×10^6 km.

SOLUTION

The semi-major axis of the orbit is $a = 60.14 \div 2$ or 30.07 A. U.
The semi-minor axis of the orbit is $b = 60.12 \div 2$ or 30.06 A. U.
The distance of each focus from the center of the ellipse is $\sqrt{a^2 - b^2}$ or 0.27 A.U.

The equation of the orbit of Neptune relative to the sun at the origin is given by

$$\frac{(x-0.27)^2}{30.07^2} + \frac{y^2}{30.06^2} = 1$$

Upon squaring and solving for y, we obtain the two equations:

$$y = \pm\sqrt{904 + 0.54x - x^2}$$

We enter these equations into the definitions for Y_1 and Y_2.
Upon setting the window to: $-32 \le x \le 32$; $-32 \le y \le 32$
and graphing these equations, we get an elliptical orbit. To see the true shape of the orbit we press: ZOOM 5 and we obtain the orbit shown in the display.

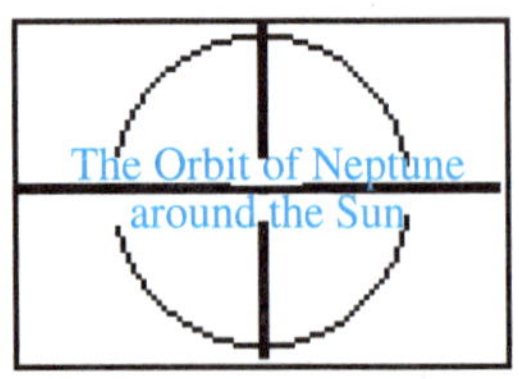

Exercises

1. a) Show algebraically that an equation of the form

$$\frac{x^2}{a^2} + \frac{y^2}{b^2} = 1$$

cannot define a function.

b) Show (geometrically) how the graph of this equation violates the definition of a function.

2. a) Write the equation of an ellipse with foci on the x-axis, center (0, 0), major axis 18 and minor axis 12.

b) What is the sum of the distances of every point on this ellipse from the two foci?

c) Graph this ellipse and use the **value** command on the CALC menu to find the coordinates of the point on the ellipse with x-coordinate 7.

3. Write the equation of the curve containing all the points whose total distance from the two points (–6, 0) and (6, 0) is 20 units. Then graph this equation.

4. For the ellipse defined by $\frac{x^2}{25} + \frac{y^2}{16} = 1$:

a) Give the coordinates of:

i) the points of intersection with the coordinate axes.

ii) both foci.

b) Solve the equation for y and graph it on your TI-84 Plus. Trace along the curve to verify the points of intersection with the x-axis and y-axis. Explain any difficulties which you encounter.

c) Try to determine the points of intersection with the x-axis by using the **intersect** command on the CALC menu to obtain the points of intersection of the two branches of the ellipse? Explain your findings.

5. Graph the equation $\frac{x^2}{16} + \frac{y^2}{25} = 1$.

a) Give the coordinates of the points where the graph intersects the coordinate axes. Compare this graph with the one in *exercise* 4 and record how they are alike and different. Determine the points of intersection of these two ellipses.

b) How could you calculate the points of intersection of the two ellipses by using the fact that either ellipse can be mapped onto the other by a reflection in the line with equation $y = x$?

6. Solve the pair of equations in *worked example 2* to find the points of intersection of the two given ellipses. Compare the answers obtained algebraically with those obtained with the **intersect** command on the TI-84.

Investigations

7. This table gives the mean distance (in A.U.s) from the sun and the orbital period (in years) of 5 planets.

a) Complete the table to test Kepler's third law.

	Mean Solar Distance R	Orbital Period P	R^3	P^2	R^3/P^2
Mercury	0.39	0.24			
Venus	0.72	0.62			
Mars	1.52	1.88			
Jupiter	5.20	11.86			
Saturn	9.55	29.46			

Explain why you might expect some of the numbers in the last column to differ from 1.

b) The mean solar distance for Neptune is about 30.1 A.U. Use this information and Kepler's third law to determine Neptune's orbital period.

8. Mars travels around the sun in an elliptical orbit with $a = 1.5230$ A. U. and $b = 1.5164$ A. U.

a) Graph the orbit of Mars around the sun relative to coordinate axes with the sun at one focus.

b) Trace around the curve to find the shortest (perihelion) and farthest (aphelion) distances to the sun. What do you discover?

9. The *eccentricity* (denoted e) of an ellipse is defined to be the ratio c/a, for an ellipse with foci ($\pm c$, 0) and semi-major axis a.

a) Describe the shape of an ellipse with $e = 0$.

b) Use the information in *exercise* 8 to determine the eccentricity of the orbit of Mars.

10. a) Write the equation of the path traced out by a point P(x, y) which moves so that its distance from the point F(c, 0) is e times its distance from the line $x = d$, if e is a constant such that $0 \le e < 1$.

b) Write an expression for d in terms of c and e for which the path becomes an ellipse with center at the origin.

c) Substitute into your equation in part a) the expression for d that you found in part b). Reduce this equation to standard form; i.e. with constant 1 on the right side.

d) Express the semi-major and semi-minor axes in terms of c and e. Is e the same as the eccentricity defined in *exercise* 9?

EXPLORATION 19 From f(x)unctions to Relations – The Hyperbola

According to legend, the oracle at Delos in ancient Greece asserted that a plague that infected Athens could only be removed by doubling the size of the cubical alter of the god, Apollo. This required construction, with ruler and compasses, of a length equal to $\sqrt[3]{2}$. About 350 B.C., the Greek mathematician Menaechmus in an attempt to solve this problem, discovered the ellipse, parabola and hyperbola. Since these curves were found to be the those that arise from the intersection of a double cone by planes of various kinds, they were called *conic sections.*

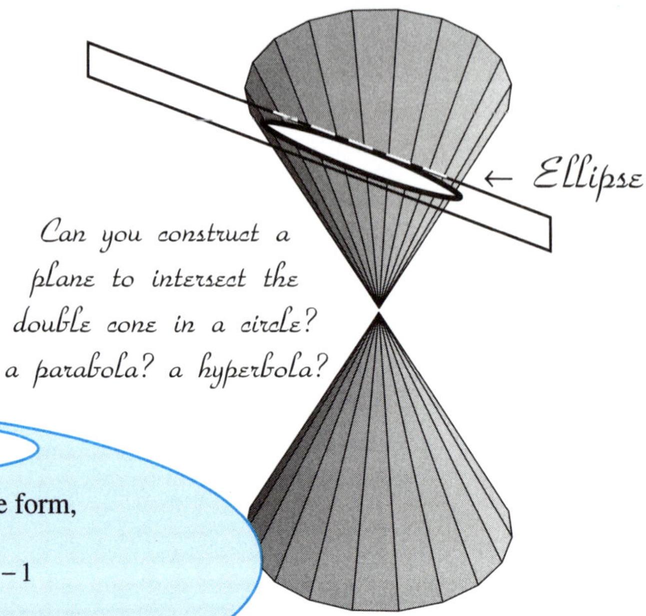

What is a Hyperbola?

Any curve defined by an equation of the form,

(i) $\dfrac{x^2}{a^2} - \dfrac{y^2}{b^2} = 1$ or (ii) $\dfrac{x^2}{b^2} - \dfrac{y^2}{a^2} = -1$

where a and b are both constants, is a *hyperbola.*

Case (i):

the *transverse* axis of the hyperbola lies along the x-axis.
the *conjugate* axis of the hyperobola lies along the y-axis.

a is the semi-transverse axis

b is the semi-minor axis

Case (ii):

the *transverse* axis of the hyperbola lies along the y-axis.
the *conjugate* axis of the hyperobola lies along the x-axis.

a is the semi-transverse axis

b is the semi-minor axis

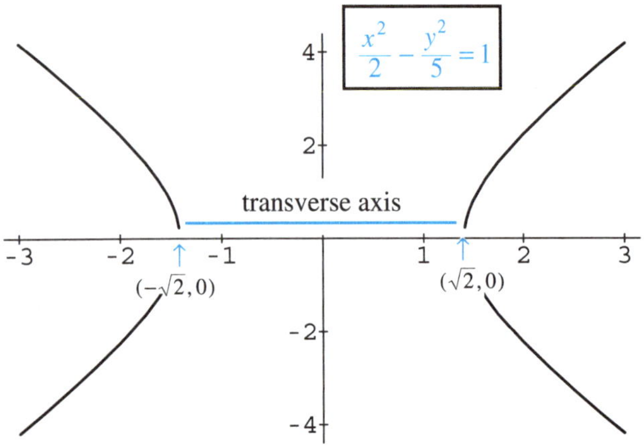

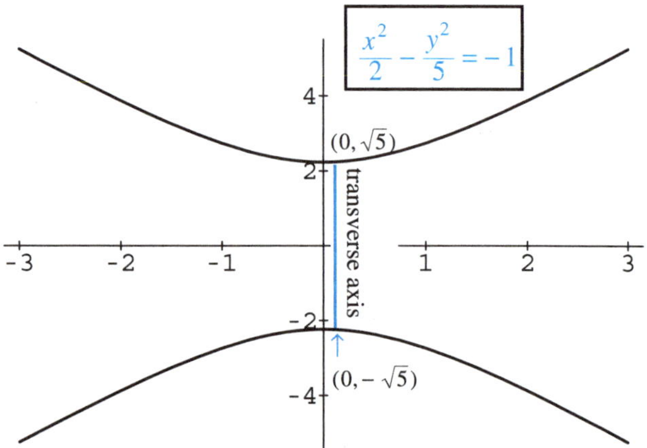

On solving the equation $\dfrac{x^2}{a^2} - \dfrac{y^2}{b^2} = 1$ for y we obtain:

$$y = \frac{b}{a}\sqrt{x^2 - a^2} \quad \text{and} \quad y = -\frac{b}{a}\sqrt{x^2 - a^2}$$

Each of these equations defines a function, one function consists of the portion of the hyperbola above the x-axis and the other function is the portion below the x-axis. Since the hyperbola has two different values of y corresponding to values of x $(x > a)$, it is not itself a function, but, like the ellipse, belongs to a more general class of sets called *relations.*

Asymptotic Behavior

As $x \to \pm\infty$, $\dfrac{b}{a}\sqrt{x^2 - a^2} \to \begin{cases} \dfrac{b}{a}x \text{ for } x \geq 0 \\ -\dfrac{b}{a}x \text{ for } x < 0 \end{cases}$

so the graph of the hyperbola approaches the graphs of the lines defined by $y = \pm\dfrac{b}{a}x$. These lines, which bound the two branches of the hyperbola, are called its *asymptotes.* The asymptotes for a particular hyperbola are shown in *worked example 2.*

WORKED EXAMPLE 1

Derive the equation of the curve traced out by a point moving so that the *difference* of its distances from the points $(-c, 0)$ and $(c, 0)$, is a constant k, $(k \neq 0)$. Identify the curve from its equation.

SOLUTION

a) If F_1 and F_2 denote the points $(-c, 0)$ and $(c, 0)$ respectively, and $P(x, y)$ is any point on the curve, then the condition that the difference in the distances is k can be written $|PF_1 - PF_2| = k$, i.e.

$$\sqrt{(x+c)^2 + y^2} - \sqrt{(x-c)^2 + y^2} = \pm k \quad ①$$

Arranging the radicals so they appear on opposite sides of equation ① yields:

$$\sqrt{(x+c)^2 + y^2} = \sqrt{(x-c)^2 + y^2} \pm k \quad ②$$

Squaring both sides of ② yields: $(x+c)^2 + y^2 = (x-c)^2 + y^2 \pm 2k\sqrt{(x-c)^2 + y^2} + k^2$ ③

Collecting terms and simplifying ③, we obtain: $4cx - k^2 = \pm 2k\sqrt{(x-c)^2 + y^2}$ ④

Squaring both sides of ④ and collecting like terms yields: $\dfrac{x^2}{\left(\frac{k}{2}\right)^2} - \dfrac{y^2}{c^2 - \left(\frac{k}{2}\right)^2} = 1$ ⑤

Equation ⑤ is of the form $\dfrac{x^2}{a^2} - \dfrac{y^2}{b^2} = 1$ with $a = k/2$ and $b = \sqrt{c^2 - a^2}$;

therefore it is a hyperbola with transverse axis the line segment from $(-k/2, 0)$ to $(k/2, 0)$.

Do we get the same equation if we interchange the points $(c, 0)$ and $(-c, 0)$?

WORKED EXAMPLE 2

a) Write the equation of the hyperbola with foci at $(-6, 0)$ and $(6, 0)$ and transverse axis 10.
b) Graph this equation and its asymptotes. State the intercepts and the conjugate axis. Give the coordinates of the points on the hyperbola with first coordinate $x = 9$.

SOLUTION

a) We substitute $c = 6$ and $k = 10$ into equation ⑤ above to obtain: $\dfrac{x^2}{25} - \dfrac{y^2}{11} = 1$

To graph this equation, we solve for y to obtain the two functions Y_1 and Y_2 corresponding to its branches above and below the x-axis.

$$Y_1 = \frac{\sqrt{11}}{5}\sqrt{x^2 - 25} \quad \text{and} \quad Y_2 = -Y_1$$

To graph the asymptotes, we note that as $x \to -\infty$ and as $x \to \infty$,

$\sqrt{x^2 - 25} \approx \sqrt{x^2}$, and so $Y_1 \approx \dfrac{\sqrt{11}}{5}x$ and $Y_2 \approx -\dfrac{\sqrt{11}}{5}x$

and so the asymptotes are defined by

$$Y_3 = -\frac{\sqrt{11}}{5}x \quad \text{and} \quad Y_4 = \frac{\sqrt{11}}{5}x$$

b) Graphing these functions in the standard window yields the graph shown in the display. The x-intercepts are ± 5 and the conjugate axis is $2b = 2\sqrt{11}$. Using the CALC menu, we evaluate Y_1 at $x = 9$ to find the point $(9, 4.963\ldots)$. By symmetry, the point $(9, -4.963\ldots)$ is on the graph of Y_2.

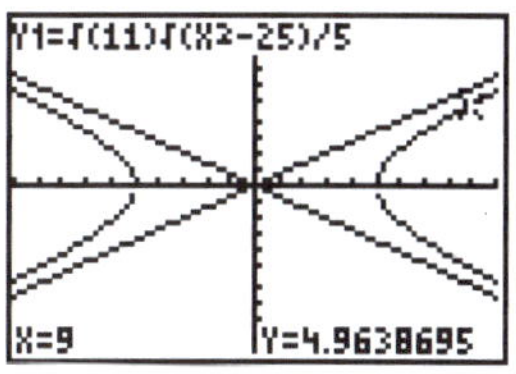

On June 26, 1993, the 24th and final satellite in the NAVSTAR global positioning system (GPS) was put in orbit around the earth. The satellites were positioned so that between 5 and 8 satellites are accessible to every location on earth at every moment of the day and night.

A radio signal from a satellite in position F_1 to a ship in position P is timed by measuring the difference between the time the signal was sent from the satellite and the time it is received by the ship. The distance PF_1 from that satellite to the ship is calculated by multiplying this time by the velocity of the radio signal.

Similarly, the distance PF_2 from P to another satellite at F_2 is recorded. The ship is then PF_1 units from F_1 and PF_2 units from F_2, so it lies at the intersection of the circles as shown.

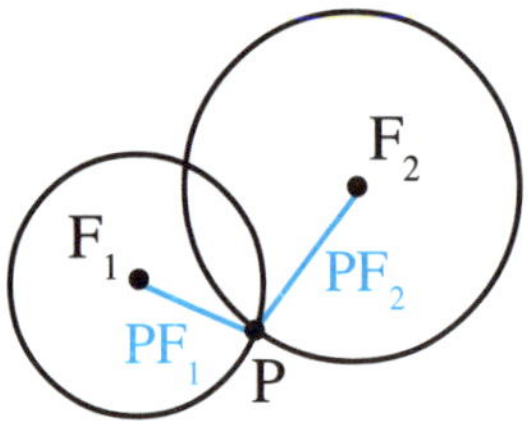

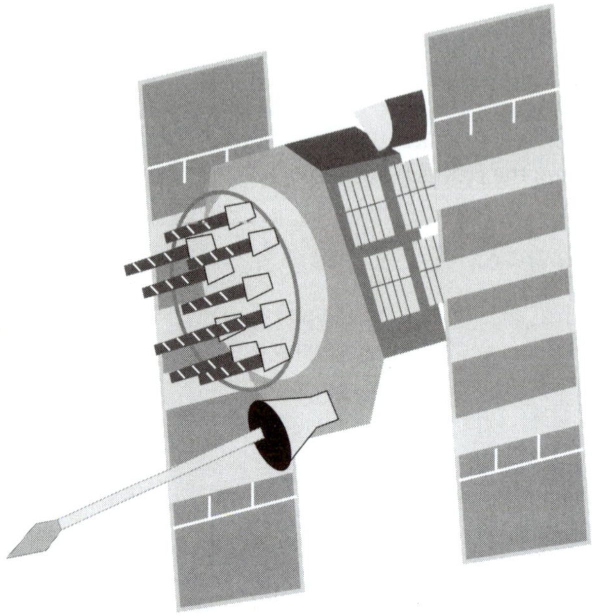

The NAVSTAR satellites (acronym for **Nav**igation **S**atellite **T**iming **and** **R**anging) constitute a Global Positioning System that enables anyone traveling anywhere on earth to locate their position instantly if they have a proper receiver.

Since radio waves travel at the speed of light (3×10^8 m/s), the smallest discrepancy in the receiver clock produces a large error e in the measurement of the distances PF_1 and PF_2. To eliminate this error, NAVSTAR computes the difference $|PF_2 - PF_1|$, which we denote k_1, and the error e drops out (see exercise 7). That is, P is located somewhere on the curve defined by $|PF_2 - PF_1| = k_1$. As seen in *worked example 2*, this is a hyperbola with foci F_1 and F_2! By computing $|PF_3 - PF_1| = k_2$ for another satellite at F_3 and finding the intersection point(s) of these two hyperbolas, NAVSTAR locates P.

WORKED EXAMPLE 3

NAVSTAR satellites are located at positions $F_1(0, 0)$, $F_2(0, 8)$ and $F_3(12, 0)$. The navigator of a ship located at an unknown position P determines that $|PF_2 - PF_1| = 6$ and $|PF_3 - PF_1| = 4$. Use this information to determine the precise coordinates of P given that P is near (3, 9).

SOLUTION

The condition $|PF_2 - PF_1| = 6$ defines a hyperbola with foci at (0, 0) and (0, 8), so $c = 4$ and $k = 6$. That is, the hyperbola has center (0, 4), $a = 3$, and $b = \sqrt{c^2 - a^2} = \sqrt{7}$.

The equation of a hyperbola with foci $(0, -c)$, $(c, 0)$ and semi-transverse axis a is: (See *exercise* 6 p.77)

$$\frac{x^2}{b^2} - \frac{y^2}{a^2} = -1 \quad \text{where } b^2 = c^2 - a^2$$

So P lies on the hyperbola with equation:

$$\frac{x^2}{7} - \frac{(y-4)^2}{9} = -1$$

Similarly, $|PF_3 - PF_1| = 4$ implies that P lies on the hyperbola with equation:

$$\frac{(x-6)^2}{4} - \frac{y^2}{32} = 1$$

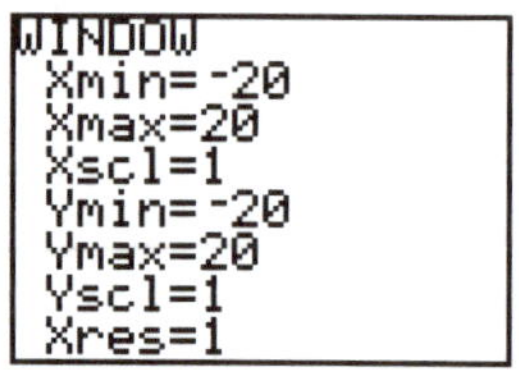

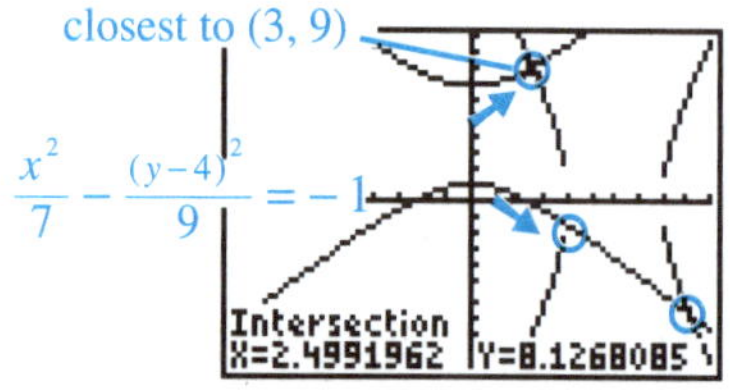

Graphing both hyperbolas in the window shown and solving for their intersections as in *worked example 2* p. 72 yields the intersection points as shown in the display. The closest intersection point to (3, 9) is (2.499…, 8.126…).

Exercises

1. a) Show algebraically that an equation of the form

$$\frac{x^2}{a^2} - \frac{y^2}{b^2} = 1$$

cannot define a function.

b) Show (geometrically) how the graph of this equation violates the definition of a function.

2. a) Write the equation of a hyperbola with foci on the x-axis, center (0, 0), transverse axis 8 and conjugate axis 12.

b) What is the difference of the distances of every point on this hyperbola from the two foci?

c) Graph this hyperbola and use the **value** command on the CALC menu to find the coordinates of the point on the hyperbola with x-coordinate 2. Explain what you discover.

3. a) Write the equation of the curve containing all the points that are 8 units farther from (–5, 0) than from (5, 0). Then graph this equation. Is this curve a function? Give reasons for your answer.

b) Write the equation of the curve containing all the points that are 8 units closer to (–5, 0) than to (5, 0). Then graph this equation. Is this curve a function? Give reasons for your answer.

c) Compare your equations in part a) and part b). What do you discover? Is there any point which satisfies both equations? Explain.

4. For the hyperbola defined by $\frac{x^2}{25} - \frac{y^2}{16} = 1$

a) Give the coordinates of:

i) the points of intersection with the coordinate axes.

ii) both foci.

b) Solve the equation for y and graph it. Use the **intersect** command on the CALC menu to obtain the points of intersection of the two branches of the hyperbola and verify your answer in part a).

c) Graph the asymptotes of this hyperbola.

5. Graph the equation $\frac{x^2}{25} - \frac{y^2}{16} = -1$.

a) Give the coordinates of the points where the graph intersects the coordinate axes. Graph its asymptotes. Compare this graph with the one in *exercise* 4 and record how they are alike and different.

b) Determine the points of intersection (if any) of these two hyperbolas. Explain your findings.

Investigations

6. a) Derive the equation of the hyperbola traced out by a point moving so that the difference of its distances from the points $(0, -c)$ and $(0, c)$ is a constant k, $(k \neq 0)$. Why do we exclude the case $k = 0$?

b) Write the equations of the asymptotes to the hyperbola defined in part a).

c) Write the equation of the hyperbola with transverse axis 6 and foci at (0, –8) and (0, 8). Graph the hyperbola and its asymptotes.

7. a) Let PF_1 and PF_2 denote the NAVSTAR measurements of the distances from P to F_1 and to F_2 respectively. Let e denote the (unknown) error in both measurements caused by the receiver clock, so that the true distances are:

$$d_1 = PF_1 + e \quad \text{and} \quad d_2 = PF_2 + e$$

Verify that $|d_1 - d_2| = |PF_1 - PF_2|$.

b) A ship at an unknown position $P(x, y)$ is somewhere in the region $10 \le x \le 15$; $-10 \le y \le -15$. NAVSTAR satellites located at $F_1(0, 0)$, $F_2(0, -8)$, and $F_3(14, 0)$ have recorded the following information.

$|PF_1 - PF_2| = 4$ and $|PF_1 - PF_3| = 6$.

Find the exact location of P.

Note: When you apply the intersect command, be sure you select the intersecting curves carefully.

8. a) Write the equation of the path traced out by a point $P(x, y)$ which moves so that its distance from the point $F(c, 0)$ is e times its distance from the line $x = d$, if e is a constant such that $e > 1$.

b) Write an expression for d in terms of c and e for which the path becomes a hyperbola with foci on the x-axis.

CHALLENGE

a) Graph the hyperbola defined by the equation

$$x^2 - y^2 = 1$$

and graph its asymptotes.

b) Express the equation of the hyperbola in part a) relative to the asymptotes as coordinate axes. That is, the first component x^* of a point is its distance from one asymptote and the second component y^* is its distance from the other asymptote.

EXPLORATION 20 Polar Coordinates

In 1671, Newton published his *Method of Fluxions*, in which he presented eight new types of coordinate systems. His seventh system is known today as *polar coordinates*. In polar coordinates, every point on an ellipse is given by an ordered pair, (r, θ), where r is the distance of the point from the origin and θ is the angle relative to the positive horizontal axis.

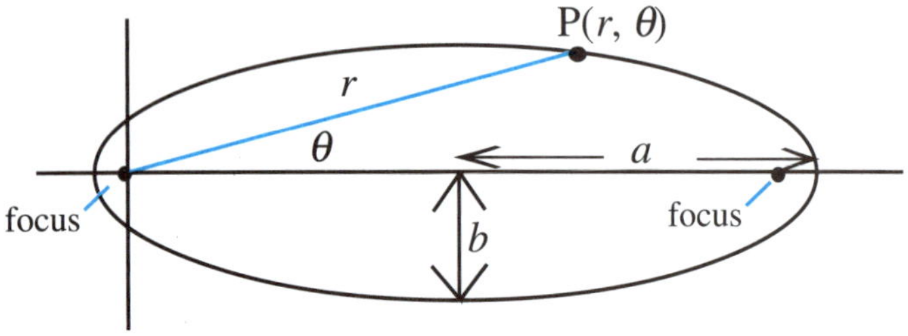

In polar coordinates, the equation of an ellipse with semi-major axis, a and semi-minor axis, b and one focus at the origin is given by the equation:

$$r = \frac{b^2}{a - \sqrt{a^2 - b^2}\cos\theta}$$

© 1997 by Sidney Harris

"WHAT AN EGO! ONE DAY IT'S NEWTON'S LAWS OF DYNAMICS, THEN IT'S NEWTON'S THEORY OF GRAVITATION, AND NEWTON'S LAW OF HYDRONAMIC RESISTANCE, AND NEWTON'S THIS AND NEWTON'S THAT."

WORKED EXAMPLE 1

a) Planet Pluto travels in an elliptical orbit with semi-major axis $a = 39.44$ AU, and semi-minor axis $b = 38.21$. Write the equation of the orbit of Pluto in polar form and then graph it.

b) Trace along your graph to determine how far Pluto is from the sun when $\theta = 0.7853\ldots$radians.

Solution

a) To graph the orbit of Pluto using polar coordinates, with the sun at the origin, we must first enter polar coordinate mode. To do this, press: **MODE** and move the cursor down three rows and over two columns. Then press **ENTER** From the mode menu also select **Radian** mode.

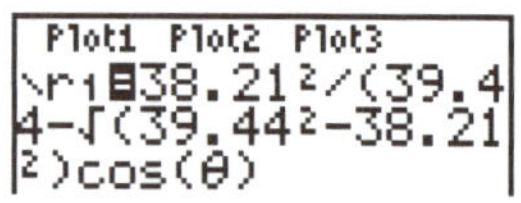

To enter the polar equation of Pluto's orbit, press **Y =** . We define r_1 by substituting $b = 38.21$ and $a = 39.44$ into the equation above to obtain the top display.

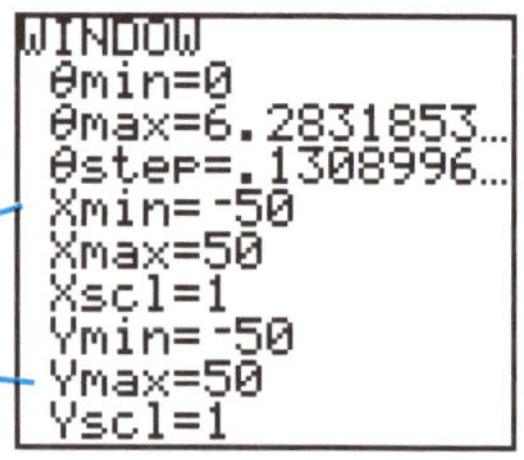

Then press **ENTER** to put r_1 into memory and use **WINDOW** to set these range values: $-50 \le x \le 50$; $-50 \le y \le 50$.
Finally we press: **GRAPH** **ZOOM** **5** obtain the graph of the orbit of Pluto shown in the lower display.

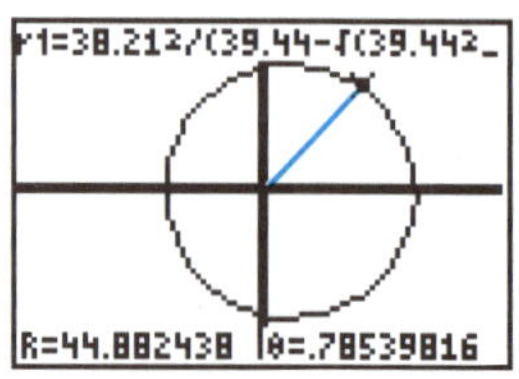

b) To display the polar coordinates during tracing, press: **2nd** **FORMAT** and select **PolarGC**, for polar graphing coordinates.
Tracing along the curve to $\theta = 0.7853\ldots$, we obtain this display indicating that $r = 44.88$ AU. That is, Pluto is 44.88 AU from the sun.

A particularly useful form of the polar equation for ellipses involves the so-called *Keplerian elements* a and e where $e = c/a$ is called the *eccentricity* (see *exercise* 9 p.73). The eccentricity of an ellipse is the measure of how much it deviates from the circular shape; an eccentricity of 0 corresponds to a circle, and eccentricities close to 1 correspond to elongated ellipses in which a is much greater than b. If we substitute $a^2 - c^2$ for b^2, and ae for c in the equation on page 78, we obtain the following special polar form of the equation of an ellipse centered at one focus and expressed in terms of a and e. (Verify this for yourself.)

$$r = \frac{a(1-e^2)}{1-e\cos\theta}$$

Space scientists often use this equation to define the orbits of planets and space vehicles. During a space mission, you can access the NASA shuttle page on the World Wide Web and you will be given the values of a and e for the current orbit. You can then enter these values of a and e into the special polar form of the equation of an ellipse to graph the orbit in polar coordinates, as shown in the following example.

WORKED EXAMPLE 2

In October 1995, the NASA shuttle page on the World Wide Web issued the Keplerian elements $a = 6\,644$ km and $e = 0.0002952$ for the Space Shuttle Mission (STS-73).

a) Graph the orbit of this space shuttle.

b) On this same graph show an outline of the earth as a circle of radius 6 378 km .

c) Trace along the orbit to find the apogee (greatest distance above the earth) and perigee (smallest distance above the earth).

Solution

a) We first select **Polar** mode and **PolarGC** format as shown in *worked example* 1.
Then we press **Y =** and define r_1 by entering the values of a and e into the formula given above, to obtain the expression shown in the display.

```
Plot1 Plot2 Plot3
\r1■6644(1-.0002
952²)/(1-.000295
2cos(θ))
```

We then set the window variables as shown and press **ZOOM** **5** to get the true shape. (Check that your TI-84 Plus is in radian mode.)

settings before ZOOM 5

```
WINDOW
 θmin=0
 θmax=6.2831853…
 θstep=.1308996…
 Xmin=-7000
 Xmax=7000
 Xscl=1
 Ymin=-7000
 Ymax=7000
 Yscl=1
```

b) To show the outline of the surface of the earth, we press **Y =** and define $r_2 = 6378$. This yields the inner ellipse when both are graphed on the same display.

c) When we press **TRACE** the cursor appears at $\theta = 0$. Since R decreases as we trace away from $\theta = 0$, the value of R must be largest at $\theta = 0$, where $R = 6645.96$. That is, the greatest height STS-73 reaches above the earth's surface is:

6645.96 – 6378 or about 268 km

Similarly, we find the height is about 264 km at perigee, where $\theta = 3.14159$ radians or 180°.

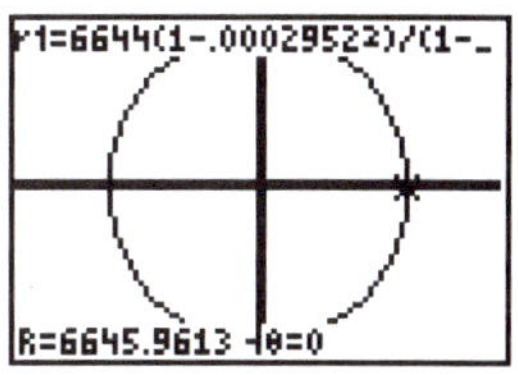

Exercises

1. The point P has polar coordinates $(5, \pi/2)$ where the first coordinate is given in meters and the second coordinate in radians. Describe the position of P relative to the origin and the polar axis (i.e. the positive x-axis).

2. a) Convert the rectangular coordinates of each point to polar coordinates, expressing the second component in degrees.

$P(3, 4)$, $Q(-5, 12)$, $R(7, -7)$ and $S(-1, -\sqrt{3})$

Are your answers unique? Explain.

b) Write a formula to express r in terms of x and y.

c) Write a formula to express θ in terms of x and y.

3. a) Convert the polar coordinates of each point to rectangular coordinates. (All second coordinates are expressed in radians.)

$A(1, \pi/4)$, $B(-5, 3\pi/2)$, $C(-4, -\pi/3)$ and $D(-1, 0.5235)$

Are your answers unique? Explain.

b) Write a formula to express x in terms of r and θ.

c) Write a formula to express y in terms of r and θ.

4. Graph the curve defined by the polar equation

$r = \theta$ where θ is in radians.

(Use the window $0 \le \theta \le 10\pi$; $-50 \le x \le 50$; $-50 \le y \le 50$.)

This curve, was discovered by Archimedes around 225 b.c.

a) Give the coordinates of the points where the curve intersects the polar axis. Explain any pattern you find.

b) Explain how the number of intersections with the polar axis is related to the upper and lower bounds on θ.

5. Use polar coordinates (polar mode) to graph each equation. After each graph is drawn, press:

ZOOM 6 followed by ZOOM 5

a) $r = 7$ b) $r = 8 \cos 3\theta$ c) $r = 5 + 5\cos \theta$

Give the coordinates of the points where each curve intersects the polar axis. Explain any pattern you find.

6. Graph each of these equations in polar mode using appropriate window settings.

a) $r = 2/(1 - \cos\theta)$ b) $r^2 = 6\cos(2\theta)$

7. Mars travels around the sun in an elliptical orbit with $a = 1.5230$ AU and $b = 1.5164$ AU.

Using polar coordinates, graph the orbit of Mars around the sun relative to coordinate axes with the sun at one focus. Trace around the curve to find the shortest (perihelion) and farthest (aphelion) distances to the sun.

Investigations

8. Define r_1 and r_2 by $r_1 = 9\cos\theta$ and $r_2 = 9\sin\theta$.

a) Use algebra to determine the polar coordinates of the points of intersections of the graphs of r_1 and r_2.

b) Graph both equations to verify your answer in a).

c) Explain any discrepancy between your answers in parts a) and b).

9. Graph each of these equations in polar mode.

Press ZOOM 5 .

a) $8\sin(\theta - \frac{\pi}{2})$ b) $8\sin\theta$ c) $8\sin(\theta + \frac{\pi}{2})$

Write an equation for r_4 whose graph, together with the graphs of a), b) and c) will form the graph shown in the display below.

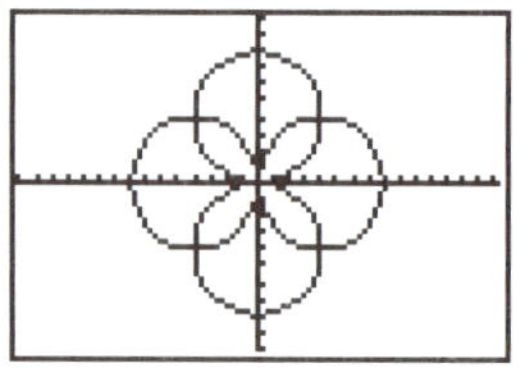

10. Assign the list { 1, 2, 3, 4 } to the variable, L_1 by entering { 1, 2, 3, 4 } STO▶ 2nd [L_1].

Then graph the equation,

$$r = 7\sin(\theta + L_1 \tfrac{\pi}{2})$$

11. Graph the following equation in polar mode.

$$r = \frac{16}{3 - \sqrt{41}\cos\theta}$$

Can you name this graph? Assign different values to the numerical coefficients, and graph those equations. Describe what you discover about the graphs of equations of this form

12. Graph the orbits of Venus, Earth and Mars in polar coordinates, given the following values of the Keplerian elements a and e where a denotes the semi-major axis in astronomical units, and e denotes the eccentricity. Set $-3 \le x \le 3$; $-2 \le y \le 2$.

Planet	a	e
Venus	0.7233	0.0068
Earth	1.0000	0.0167
Mars	1.5230	0.0934

Can you graph Pluto on this display? Why?

EXPLORATION 21 Parametric Equations

To this point, all graphs in the plane have been defined by equations in two variables. When cartesian coordinates were used, the graph has been defined by an equation in x and y. When polar coordinates were used, the graph has been defined by an equation in r and θ. Sometimes it is useful to express a relationship between x and y, (or r and θ) in terms of a third variable, called a *parameter*. For example, the ellipse defined by $x^2 + 4y^2 = 4$, could be defined by the two *parametric* equations:

$$\left.\begin{aligned} x &= 2T \\ y &= \pm\sqrt{1-T^2} \end{aligned}\right\} \Leftarrow$$

Substitute these expressions for x and y into $x^2 + 4y^2 = 4$ to verify that these parametric equations are equivalent to the cartesian equation.

It could also be defined by the parametric equations:

$$\left.\begin{aligned} x &= 2\cos T \\ y &= \sin T \end{aligned}\right\} \Leftarrow$$

© 1997 by Sidney Harris

There are an unlimited number of ways of defining any given curve using parametric equations. In *Exploration* 8, you studied the trajectory of the longest home run in professional baseball. That trajectory was defined by the equation:

$$y = 0.9x - 0.0014x^2.$$

The following example shows how that same curve can be defined by two parametric equations.

WORKED EXAMPLE 1

The trajectory of the longest hit baseball in major-league history is described by the parametric equations:

$$x = 106.9\text{T}$$
$$y = 96.2\text{T} - 16\text{T}^2$$

where T is the time in seconds from the moment of impact, x is the horizontal distance (in feet) traveled by the ball and y is its height above the ground (in feet) T seconds after impact.

a) Graph the trajectory of the ball

b) How many seconds did it take the ball to reach its maximum height?

c) What was the height of the ball 4 seconds after impact and what was the horizontal distance traveled by the ball at this time?

WORKED EXAMPLE 1 (cont'd)

Solution

a) To access parametric mode we press: [MODE] [▼] [▼] [▼] [▶] [ENTER]

To enter the equations we press: [Y=] [1] [0] [6] [.] [9] [X,T,θ,n] [ENTER]

and [9] [6] [.] [2] [X,T,θ,n] [–] [1] [6] [X,T,θ,n] [x^2] [ENTER]

From *Exploration* 8, we know that $x < 700$, $y < 460$, so we press [WINDOW] and use these settings: Xmin = 0, Xmax = 700; Ymin = 0, Ymax = 460 .

We know from experience that it takes less than 10 seconds for a ball to escape the stadium, so we set Tmin = 0, Tmax = 10 and Tstep = .1 .

When we press [GRAPH] , we obtain the same trajectory as in *Exploration* 8.

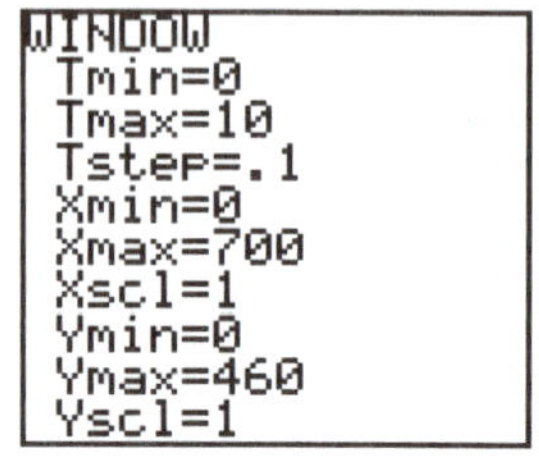

b) To approximate the time to maximum height, we press [TRACE].

We trace along the curve to the highest point in the trajectory. It seems to occur at T = 3, X = 320.7 and Y = 144.6, as shown in the display. We zoom in by pressing [ZOOM] [2] [ENTER] . However, we get no improvement in our estimate because Tstep is too large. That is, the graph is plotted for increments of 0.1 in T and the maximum occurs between two multiples of 0.1. To create smaller increments, we set Tstep = 0.01. This would plot our graph for increments of 0.01 in T. To reduce the length of time required to plot so many points, we graph the curve only in the interval of interest. That is, we set Tmin = 2.9 and Tmax = 3.1 and we leave the X and Y settings where they were after the zoom in. (Xmin = 232.71…, Xmax = 407.71…; Ymin = 83.46…, Ymax = 198.46…)

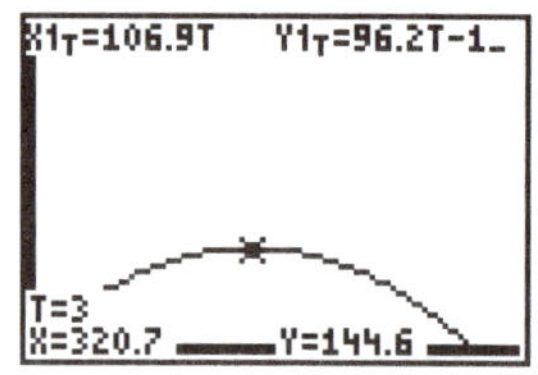

Note: Make sure you've chosen RectGC not Polar GC from the Format menu.

Upon pressing [GRAPH] we obtain the segment of the curve between $T = 2.9$ and $T = 3.1$. Tracing along the curve, we find the maximum height occurs near $T = 3.01$. (See the display.) That is, the ball reached its maximum height about 3 seconds after impact.

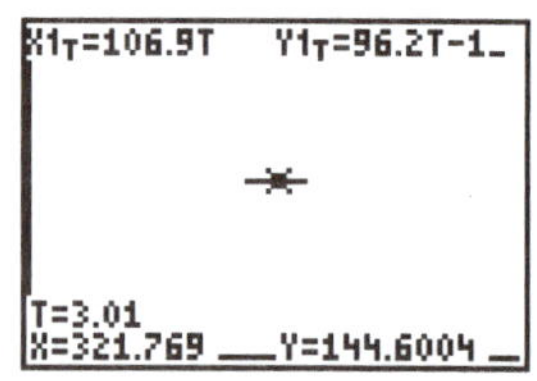

c) To determine the height and horizontal displacement 4 seconds after impact, we could substitute $T = 4$ directly into the equations. We obtain a height of 128.8 feet and a horizontal displacement of 427.6 feet. Alternatively, we could press: [2nd] [CALC] [ENTER] , then press [4] [ENTER] in response to the prompt **T=**. We obtain the same values for X and Y as by direct computation. (See bottom display.)

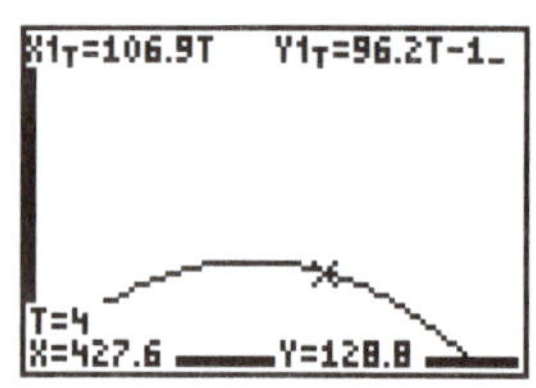

Much of what what we know about stars has come from the study of pairs of stars which orbit around each other. When the orbit is edgewise to the line of sight from earth, each star in turn eclipses the other. Such star systems are called *eclipsing binaries*. The following example shows how we can simulate the actual motion and the observed motion of such systems using parametric equations.

WORKED EXAMPLE 2

A companion star in an eclipsing binary system revolves around the larger star in an elliptical orbit given by the parametric equations:

$$x = 26\cos\frac{T}{2}$$

$$y = 20\sin\frac{T}{2}$$

where x and y are expressed in millions of kilometers and T is time measured in hours. The orbit is seen edgewise from earth so that only the y-component of the motion is in the line of sight.

a) Graph the actual orbit of the companion start and the component of its motion perpendicular to the line of sight.
b) Use the trace key to simulate the motion perpendicular to the line of sight.

Solution

a) To access parametric mode we press: [MODE] [▼] [▼] [▼] [▶] [ENTER]

To enter the parametric equations of the actual orbit, we press:

[Y =] [2] [6] [COS] [X,T,θ,n] [÷] [2] [)] [ENTER]

[Y =] [2] [0] [SIN] [X,T,θ,n] [÷] [2] [)] [ENTER]

The component of motion perpendicular to the line of sight is the y-component. To project this along the vertical line, $x = -26$, we define $X_{2T} = -26$ and $Y_{2T} = Y_{1T}$ by pressing:

[(–)] [2] [6] [ENTER] and [VARS] [▶] [2] [2] [ENTER]

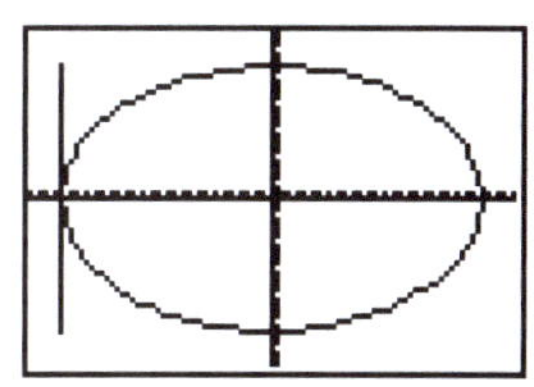

Since the orbit is an ellipse with semi-major axis 26, semi-minor axis 20, we press [WINDOW] and set: **Xmin = -30, Xmax = 30; Ymin = -25, Ymax = 25**. For a complete period to occur, the argument, T/2, must run through 2π radians, so we set: **Tmin = 0, Tmax = 4π and Tstep = .1** . Then we press [GRAPH] to obtain the graph shown in the display. To see the graph of the orbit and its projection in the vertical plane plotted simultaneously, press:

[MODE] [▼] [▼] [▼] [▼] [▼] [▶] [ENTER] [GRAPH]

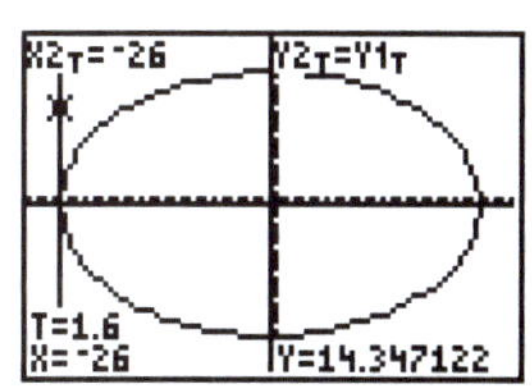

b) To place the cursor on the vertical line, press [TRACE] [▼] , then hold down the [▶] key to watch the cursor move up and down the vertical line perpendicular to the line of sight. This simulates the motion of the companion star. Though the stars are too far away to be seen individually, the fluctuating light from the eclipsing binaries enables us to verify this motion.

Exercises

1. Write each pair of parametric equations as a single equation in x and y.

a) $x = T$
$y = 2T + 1$

b) $x = 2\sin T$
$y = 2\cos T$

2. The trajectory of a golf ball is expressed in terms of the number of seconds, T, since impact by the parametric equations (where x and y are expressed in meters):

$$x = 49T$$
$$y = 20T - 4.9T^2$$

a) Express this trajectory as an equation in x and y.
b) Graph the parametric equations and determine the value of T when the golf ball reaches its maximum height.
c) How long is the golf ball in the air?

3. a) Graph the curve defined by these equations for the interval $0 \leq T \leq 2\pi$, where T is in radians.

$$x = 4\cos T$$
$$y = 3\sin T$$

b) What is the name of this curve?
c) Write these two equations as a single equation in x and y.
d) At what points does this curve intersect the coordinate axes?

4. Write the parametric equations which define each of the ellipses shown below.

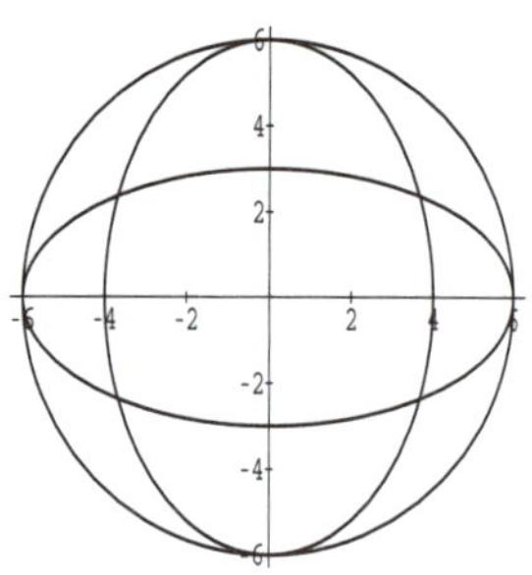

5. Graph the parametric equations for $0 \leq T \leq 4\pi$:

$$x = T\cos T$$
$$y = T\sin T$$

a) Name the curve on your display.
b) Can you express these equations as a single equation in x and y? Explain.

Investigations

6. Graph the curve defined by these equations.

$$x = 3\sec\theta \qquad 0 \leq \theta \leq 2\pi$$
$$y = 5\tan\theta$$

a) Identify the curve you have graphed.
b) State the points of intersection with the coordinate axes.
c) Write these parametric equations as a single equation in x and y.

7. Graph the curve defined by each set of parametric equations for $0 \leq T \leq 2\pi$.

a) $x = \cos T$
$y = 4\sin^2 T$

b) $x = \frac{1}{2}T$
$y = 4 - T^2$

Write an equation in x and y corresponding to each pair of parametric equations. Do both pairs of equations define the same curve? Explain why or why not.

8. Graph the curve defined by these equations

$$x = \sin 2T$$
$$y = \sin 3T$$

for the following window values: $0 \leq T \leq 2\pi$, $0 \leq x \leq 2$ and $0 \leq y \leq 2$. In how many points does this curve intersect itself? Find the values of T at each intersection point.

CHALLENGE

The position of the moon in its orbit around the earth is given by the equations: $x = 384000\cos 0.01T$

$y = 384000\sin 0.01T$

where x and y are expressed in kilometers and T is in hours. The path of a satellite, launched toward the moon is given by the equations: $x = 502T$

$y = 2258T - 4.9T^2$

a) Graph both pairs of equations using simultaneous mode and the window settings: $0 \leq T \leq 700$; Tstep = 4 and $-400000 \leq x \leq 400000$; $-400000 \leq y \leq 400000$.
b) Does the satellite collide with the moon? If not, determine when the two bodies are closest.

Answers to Exercises & Hints for Investigations

"Whatever happened to elegant solutions?"

Answers to Exercises & Hints for Investigations

Exploration 1

1. a) Let $S = 9^1 + 9^2 + 9^3 + \ldots 9^n$

Then $9S = 9^2 + 9^3 + \ldots 9^n + 9^{n+1}$.

Subtracting, we find $8S = 9^{n+1} - 9$, so $S = \dfrac{9^{n+1}-9}{8}$.

b) Let $S = a + ar + ar^2 + \ldots ar^n$.

Then $rS = ar + ar^2 + \ldots ar^n + ar^{n+1}$.

Subtracting, we find $(r-1)S = ar^{n+1} - a$, so $S = \dfrac{a(r^{n+1}-1)}{r-1}$

2. a) To generate this table, we define **Y₁ = 9^X** and **Y₂ = (9^(X+1)-9)/8** and press [2nd] [TABLE]. We obtain this display.

X	Y1	Y2
1	9	9
2	81	90
3	729	819
4	6561	7380
5	59049	66429
6	531441	597870
7	4.78E6	5.38E6
X=1		

b) From the table, we observe that $9^6 = 531441$ and the sum $9^1 + 9^2 + 9^3 + 9^4 + 9^5 + 9^6 = 597870$

c) The sums of the digits of the first 6 positive powers of 9 are respectively, 9, 9, 18, 18, 27 and 18. That is, the sum of the digits of a power of 9 is a multiple of 9, for each of the first 6 powers of 9. It can be proved that this is true for all powers of 9.

3. To generate this table, we define **Y₁ = 2^X**. When we press [2nd] [TABLE], we obtain the table shown in the display. Upon scrolling down, we obtain the three displays shown below.

X	Y1
1	2
2	4
3	8
4	16
5	32
6	64
7	128
X=7	

X	Y1
19	524288
20	1.05E6
21	2.1E6
22	4.19E6
23	8.39E6
24	1.68E7
25	3.36E7
X=20	

X	Y1
24	1.68E7
25	3.36E7
26	6.71E7
27	1.34E8
28	2.68E8
29	5.37E8
30	1.07E9
X=30	

X	Y1
34	1.7E10
35	3.4E10
36	6.9E10
37	1.4E11
38	2.7E11
39	5.5E11
40	1.1E12
X=40	

From these tables, we obtain these values of n.

a) $n = 20$ b) $n = 30$ a) $n = 40$

4. We define **Y₁ = 2^(X- 1)**, so that **Y₁** gives the number of grains on the Xth square. Define **Y₂ = 2^X-1**, the total number of grains on all squares up to the Xth. This generates the table shown on the right.

X	Y1	Y2
19	262144	524287
20	524288	1.05E6
21	1.05E6	2.1E6
22	2.1E6	4.19E6
23	4.19E6	8.39E6
24	8.39E6	1.68E7
25	1.68E7	3.36E7
X=21		

a) The table shows that the 21st square is the first one on which the number of grains exceeds 1 million and the 31st square is the first one on which the number of grains exceeds one billion.

b) The first square for which the cumulative total exceeds 1 trillion (10^{12}) is the 40th.

c) The total number of grains on all 64 squares is found by setting the TableMin = 60 and observing at X = 64, the cumulative total is 1.8×10^{19}.

Enter 2^64-1 to get 18 446 744 07E19.

Exploration 1 (cont'd)

5. a) To generate this table, we define **Y₁ = 9^X** and **Y₂=7^(X+5)** and press [2nd] [TABLE].

We scroll down to the part of the table shown in the figure. Observe from rows 38 and 39, that $9^{38} < 7^{44}$ but $9^{39} > 7^{44}$, so $9^{38} < 7^{44} < 9^{39}$, i.e. the inequality holds for n = 39.

X	Y1	Y2
35	2.5E33	6.4E33
36	2.3E34	4.5E34
37	2E35	3.1E35
38	1.8E36	2.2E36
39	1.6E37	1.5E37
40	1.5E38	1.1E38
41	1.3E39	7.5E38
X=39		

b) We define **Y₁ = 9^X/7^(X+5)** and press [2nd] [TABLE] to obtain the display shown here. The first time **Y₁** exceeds 1 occurs when **X** = 39, so this is the smallest value of n for which the inequality holds.

X	Y1
34	.30579
35	.39316
36	.50549
37	.64991
38	.8356
39	1.0743
40	1.3813
X=39	

6. a)

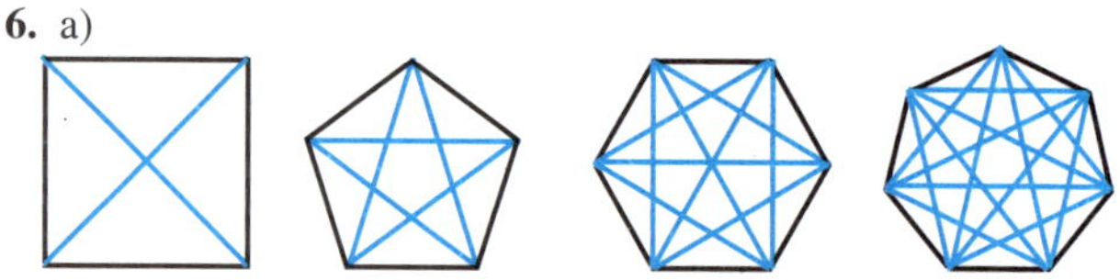

sides	4	5	6	7
diagonals	2	5	9	14

b) To determine the number of diagonals in an octagon, we observe that there are 5 diagonals from each vertex. Since there are 8 vertices, we would count 5 × 8 or 40 diagonals; however, we've counted each diagonal from both ends so there are actually 20 diagonals.

Number of Sides	Number of Diagonals
4	2
5	5
6	9
7	14
8	20

c) The sequence displayed in the second column is 2, 5, 9, 14, 20 .... We observe that the differences in successive terms are 3, 4, 5, and 6. Therefore the number of diagonals in a polygon of n sides is the number in a polygon of 4 sides plus the sum of the first $(n-4)$ terms in the series 3, 4, 5, 6, …

The sum of the first $(n-4)$ terms in the series 3, 4, 5, 6, … is 3 less than the sum of the first $(n-2)$ natural numbers, 1, 2, 3,…

The sum of the first $n-2$ natural numbers is $(n-2)(n-1)/2$ so the number of diagonals in the polygon of n sides is:

$(n-2)(n-1)/2 - 3 + 2$ which simplifies to $n(n-3)/2$.

d) To create a table that gives the number of diagonals for a polygon of n sides, we define $Y_1 = x(x-3)/2$. Then we access [2nd][TBL SET] and we set Tblstart = 4 and ΔTbl = 1. We obtain the table shown in the display.

X	Y1
4	2
5	5
6	9
7	14
8	20
9	27
10	35
X=10	

e) Scrolling down the table to X = 12 shows that the dodecagon has 54 diagonals. To count these in the diagram, we count 9 diagonals from each of 12 vertices and divide by 2 to obtain 9 × 12/2 or 54.

X	Y1
11	44
12	54
X=12	

Answers to Exercises & Hints for Investigations

Exploration 2

1. a) 0.75 m^3, 1.5 m^3, 1.5 m^3

b) slope CD = 0. This tells us that the volume of water in the tub is not changing, so either the taps are turned off or water from the taps is entering at the same rate as it is draining from the tub. Probably the water is turned off.

c) slopeAB = 0.075 slopeBC = 0.15 slopeDE = -0.3 The slope of AB indicates that water flowed at 0.075 m^3/min. in the first 10 minutes. The slope of BC indicates that the rate of flow doubled to 0.15 m^3/min during the next 5 minutes. (Probably both taps were turned on.) The slope of DE is negative, indicating that there was a decrease in the volume of water in the tub at the rate of 0.3 m^3/min during the final 5 minutes. Probably Archimedes pulled the plug.

d) Archimedes appears to have had a hot water tap, since the rate of flow doubled between points B and C. The real question should be "Is the Archimedes referred to in this question *the* Archimedes from ancient Greece".

e) There would be an instantaneous increase (decrease) in the height of the water when Archimedes entered (left) the tub.

2. Since $u(1) = -2$, we enter -2 opposite $u(n\text{Min})$ as shown in the left display. We obtain the graph shown in the right display. Tracing yields $u(8) = 40$.

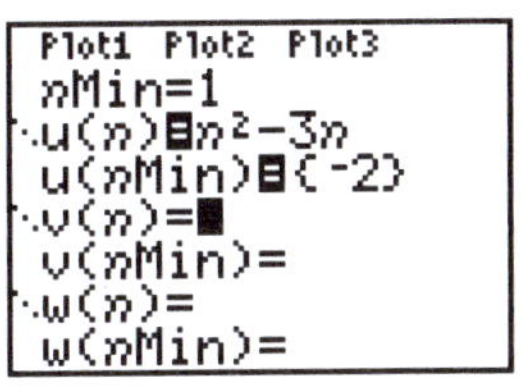

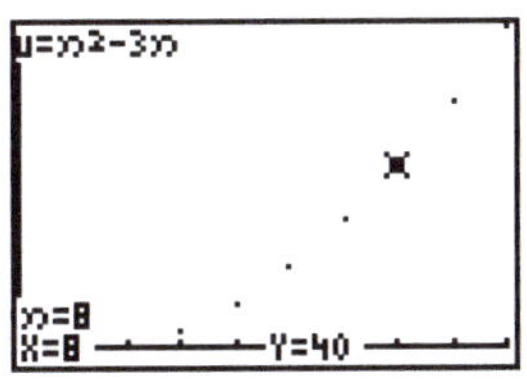

3. The graphs are compared in the displays below.

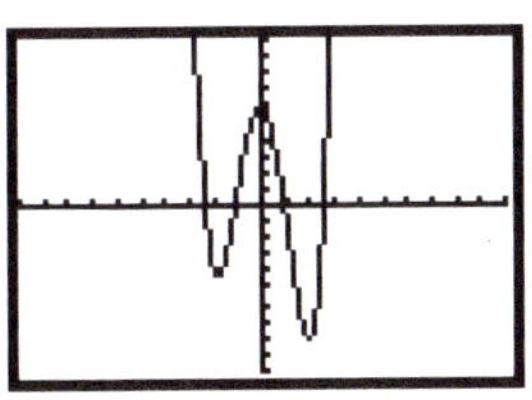

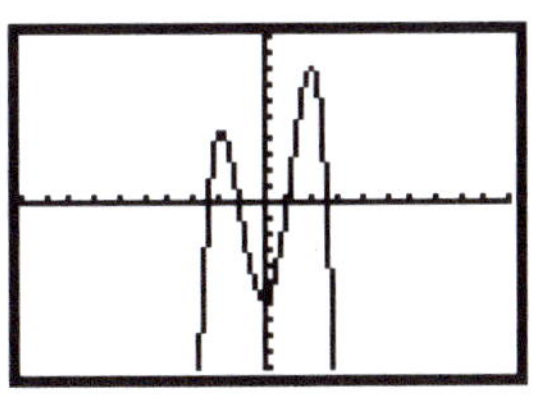

We observe that each graph is the image of the other under a reflection in the x-axis. Display both curves at the same time to obtain an interesting image.

Note

When you trace along a graph to find roots, you may observe that (owing to the pixel structure of the screen) you will obtain approximate values whose accuracy will depend somewhat on the viewing box. A method for finding the exact roots will be shown in *Exploration* 4

Exploration 2 (cont'd)

4. The graphs are as shown below for the default window settings..

a)

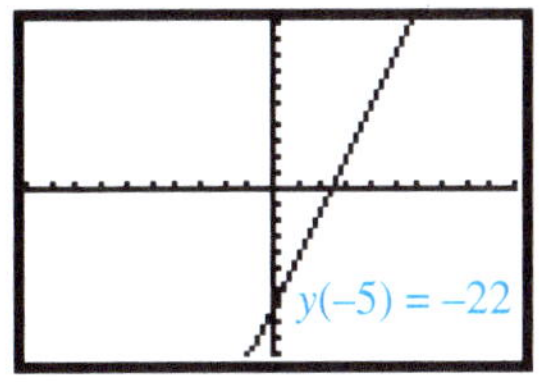

b)

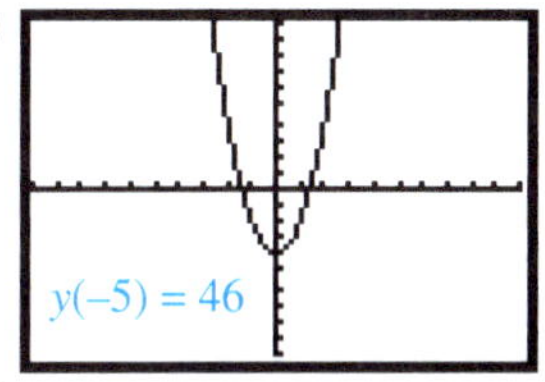

c)

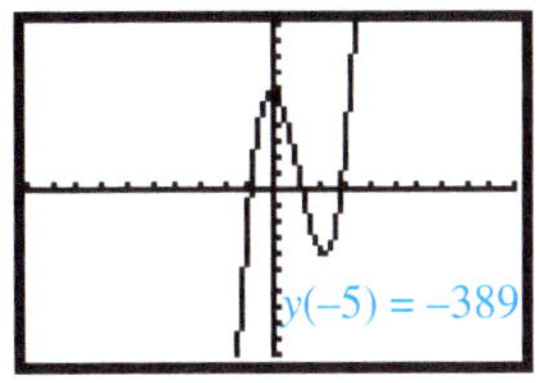

d)

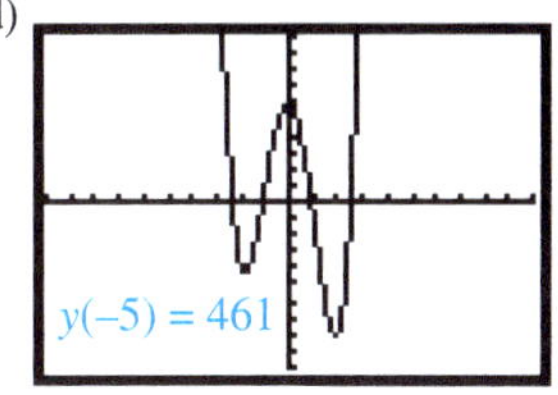

e)

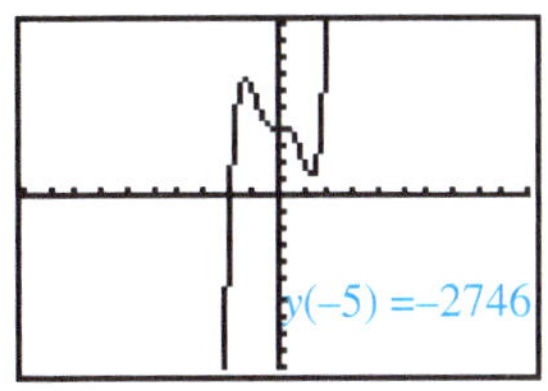

f)

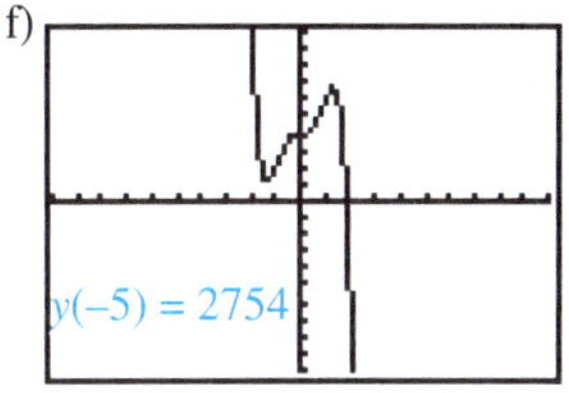

Graphs e) and f) are reflections of each other in the y-axis because their defining equations are related by the transformation $x \rightarrow -x$

5. a) With the default settings we get the graph shown below left.

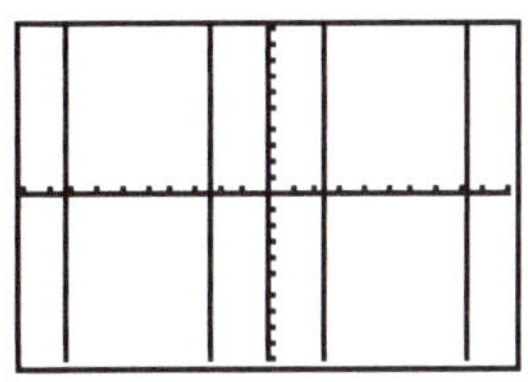

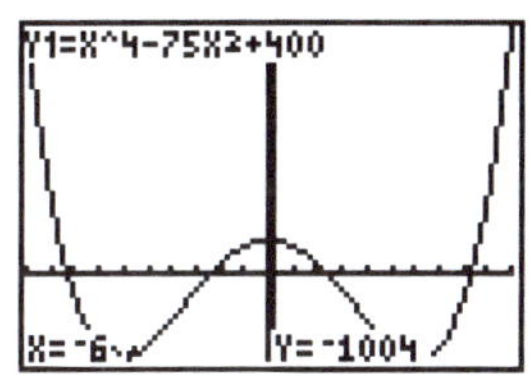

b) With the ZoomFit settings, we obtain the graph above right.

c) As shown in the diagram above, $Y_1 = -1004$ when $x = -6$.

d) The graph is symmetrical about the y-axis because the equation is unchanged under the transformation $x \rightarrow -x$.
Therefore $Y_1 = -1004$ at $x = 6$.

e) As the absolute value of the coefficient of the x^2 term is decreased the hump in the center of the graph is flattened.

6. c) When we follow the instructions by moving right and then down to create the viewing box, we obtain the display below left.

d) When we construct a viewing box by moving the cursor to (–5.07…, 761.506…) then right and down to (8.32…, –969.19…) we obtain the display below right. The approximate roots obtained by tracing are: –4.07, –1.08, 1.06, 4.05, 7.04. The actual roots are respectively –4, –1, 1, 4 and 7.

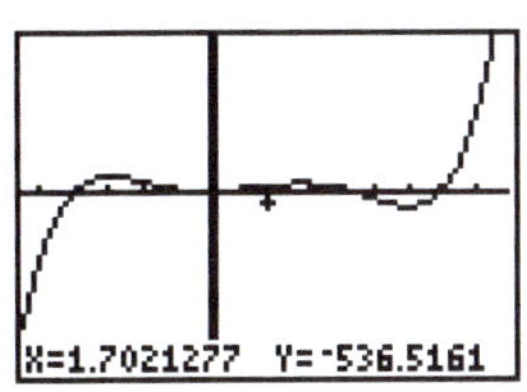

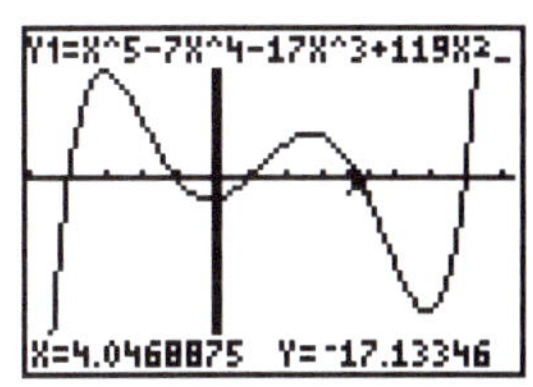

Answers to Exercises & Hints for Investigations

Exploration 3

1. a & b) c & d)

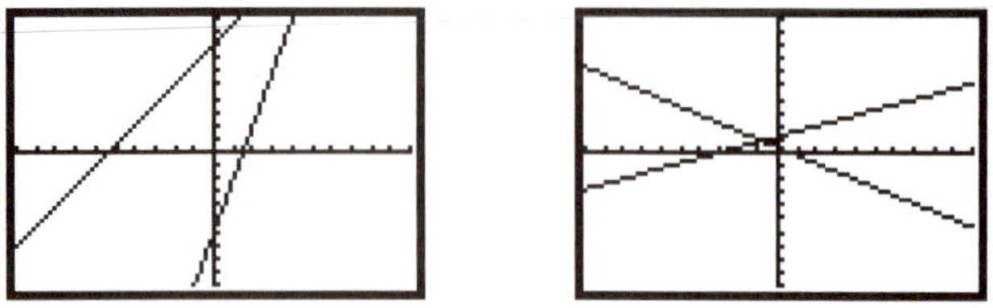

Note: In parts (c) and (d) you must first solve for y before you can enter the equation in x.

2. When the equation is expressed in the form $y = mx + b$, the value of m is the slope and the value of b is the y-intercept.
a) slope 4, y-intercept –6 b) slope 1.4, y-intercept 8
c) slope 2/5, y-intercept 6/5 d) slope –3/5, y-intercept 2/5

3. Tracing along the line until we reach the smallest value of $|y|$, we obtain these first approximations to the roots:
a) $x \approx 1.49$ b) $x \approx -5.74$ c) $x \approx -2.98$ d) $x \approx 0.67$

We can obtain approximations closer to the exact roots in exercise 3 by zooming in and tracing to the root again.

4. a) $x = 1.5$ b) $x = -40/7$ c) $x = -3$ d) $x = 2/3$

5. A linear equation has a root iff (if and only if) its graph crosses the x-axis. The only straight lines which do not intersect the x-axis are those which are parallel to it. Therefore a linear equation $Ax + By + C = 0$ has a root iff $A \neq 0$ or in the degenerate case when A =0 and C = 0.

6. The completed table is:

Equation of the Line	Slope of the Line	Equation of Inverse	Slope of the Inverse Line
$y = 3x + 7$	3	$x = 3y + 7$	1/3
$y = 3x - 7$	3	$x = 3y - 7$	1/3
$3x + 2y = 6$	–3/2	$3y + 2x = 6$	–2/3
$8x - 2y + 7 = 0$	4	$8y - 2x + 7 = 0$	1/4

We observe that the slope of the graph of a linear function and the slope of the graph of its inverse are reciprocals.

7. The equations are:
a) $4x + 11y = 50$ b) $5x - 2y = 1$ c) $x + y = 0$ d) $x - y = 0$

8. b) The display on the left shows that 20°C = 68°F. Similarly, we find that –10°C = 14°F.

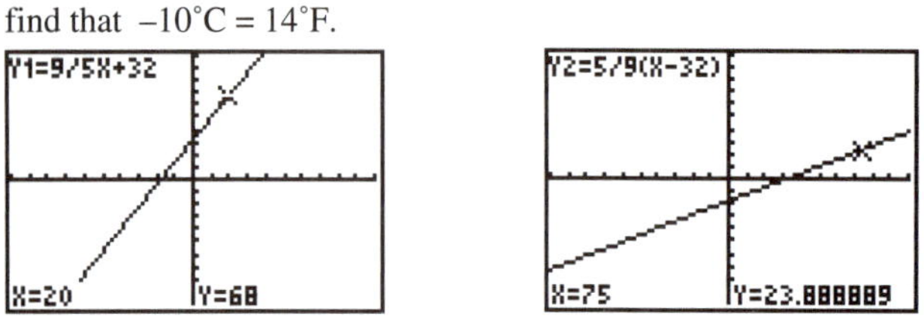

c) The display on the right shows that 75°F = 23.89°C. Similarly we find 0°F = –17.78°C.

d) The graphs intersect at (–40, –40) where both scales are the same.

Exploration 3 (cont'd)

9. The straight line is the only graph that has a constant slope. That is, the slope of the line segment joining any two points on the line is the same for all pairs of points on the line.

10. The graphs of the lines defined by equations $y = 2x$ and $y = 2x - 7$ are parallel (i.e., they both have a slope of 2) but different y-intercepts.

11. The graphs of the lines defined by equations $y = 3/2x + 5$ and $y = 3/2x - 3$ are parallel (i.e., they both have a slope of 3/2) but different y-intercepts.

12. When the equation of a linear function is written in the form $y = mx + b$, the value of m is the slope of the graph and the value of b is the y-intercept.

CHALLENGE

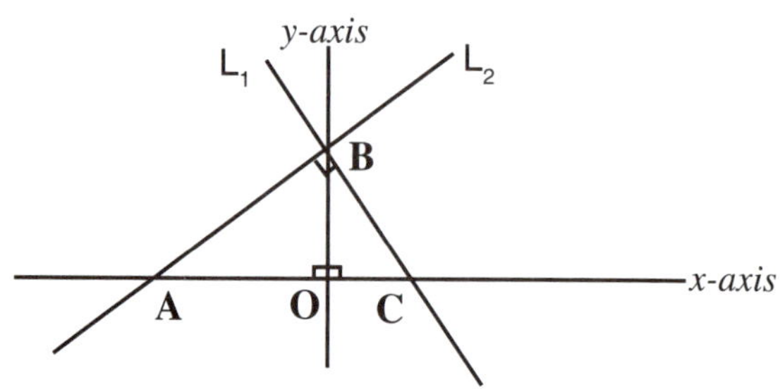

Since $\angle BAO + \angle ABO = 90^\circ$ and $\angle CBO + \angle ABO = 90^\circ$, then $\angle BAO = \angle CBO$.

Therefore ΔBAO and ΔCBO are equiangular and therefore similar. Since these triangles are similar, their ratio of the lengths of any two sides of one triangle is the same as the ratio of the corresponding sides of the other triangle. That is,

$$OB/AO = OC/OB \text{ or slope of } L_2 = -1/\text{slope } L_1.$$

We say that the slopes of perpendicular lines are negative reciprocals. Conversely if the slopes of two lines are negative reciprocals, then those two lines are perpendicular.

Investigation: Linear Functions

1. The cost as a function of the number of guests n is:
a) Leisure Lodge: $Y = 500 + 12(n - 25)$ $n > 25$
b) Cedar Glen Country Club: $Y = 14n + 90$
c) Mariners' Inn: $Y = 17n$ $n > 23$

3. The Cedar Glen and Mariners' Inn lines intersect at the point (30, 510).
The Leisure Lodge and Mariners' Inn lines intersect at the point (40, 680).
The Leisure Lodge and the Cedar Glen lines intersect at the point (55, 860).

The intersection point (n, m) gives the number of guests n at which both cost functions take the same value, m dollars.

Answers to Exercises & Hints for Investigations

Exploration 4

1. a) vertex (2, –25); axis of symmetry $x = 2$
zeros: (–3, 0) & (7, 0) range: $-25 \le y < \infty$

b) vertex (–2.75, –3.375); axis of symmetry $x = -2.75$
zeros: (–3.5, 0) & (–2, 0) range: $-3.375 \le y < \infty$

c) vertex (–1, –6); axis of symmetry $x = -1$
zeros: (–3.449, 0) & (1.449, 0) range: $-6 \le y < \infty$

d) vertex (–0.2, 7.2); axis of symmetry $x = -0.2$
zeros: (–1.4, 0) & (1, 0) range: $-\infty < y \le 7.2$

2. The formula yields the following pairs of roots:

a) $x = -3$ and 7 b) $x = -2$ and $-7/2$

c) $x = -1 - \sqrt{6}$ and $-1 + \sqrt{6}$ d) $x = -1.4$ and 1

We observe that the zeros of the functions in *exercise* 1 are the same as the roots of the corresponding equations, however, the TI-84 Plus gives decimal representations while the formula yields fractions as common fractions and roots in "surd" notation.

3. The graphs are shown in the displays below with the zeros given when they exist (i.e., when they are real).

a)

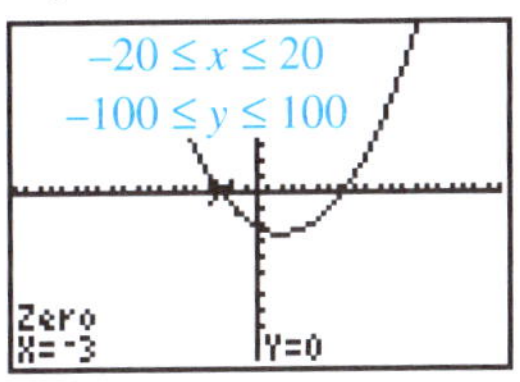

zeros:
(– 3, 0); (7, 0)

b)

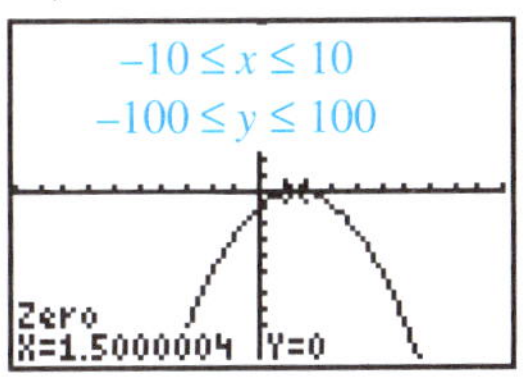

zeros:
(1.5, 0); (1.5, 0)

c)

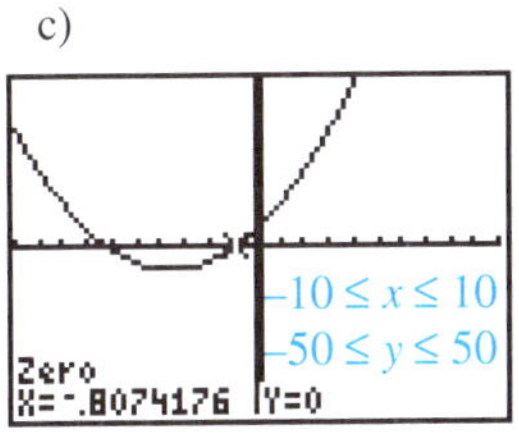

zeros:
(– 6.19…, 0); (–0.807…, 0)

d)

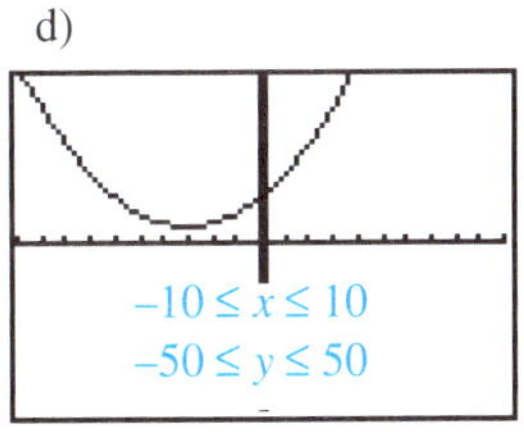

zeros:
There are no real zeros

The graph in 3d) yields an error message when we attempt to find zeros because the graph does not intersect the x-axis, i.e., there are no real zeros of this function and therefore no real roots of the corresponding equation.

The number of real zeros of a function is the number of points of intersection it has with the x-axis, but this is also the number of real roots of the corresponding equation, so there is the same number of each.

Exploration 4 (cont'd)

4. The diagram shows that the golden rectangle was constructed by bisecting the base of a square measuring 2 units per side. Then an arc is described with center B, the midpoint of the base and radius the distance to D of the square. The point C where the arc cuts the extended base of the square is then taken to be the bottom right corner of the golden rectangle.

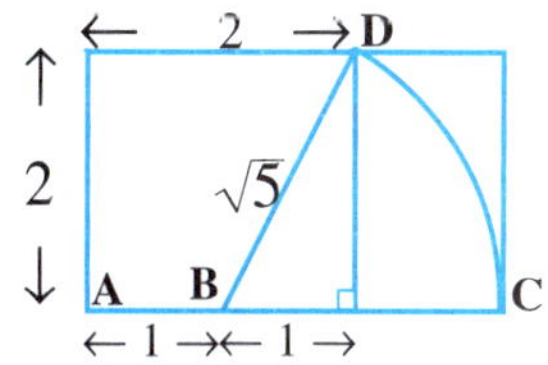

It follows from the Pythagorean Theorem that BD has length $\sqrt{5}$. Therefore the length of AC is $1 + \sqrt{5}$ and the ratio of the length to the width of the rectangle is $(1 + \sqrt{5})/2$, the golden ratio.

5. a) $\sigma \times \tau = \left(\frac{1-\sqrt{5}}{2}\right)\left(\frac{1+\sqrt{5}}{2}\right) = \frac{1-(\sqrt{5})^2}{4} = -1$ so $\sigma = -1/\tau$

b) $\tau^2 = \frac{3+\sqrt{5}}{2}$, $\tau^3 = 2+\sqrt{5}$, $\tau^4 = \frac{7+3\sqrt{5}}{2}$, $\tau^5 = \frac{11+5\sqrt{5}}{2}$, …

c) We see that $\tau^{n+1} = \tau^n + \tau^{n-1}$ so, $\tau^6 = \tau^5 + \tau^4$.
Therefore, $\tau^6 = 9 + 4\sqrt{5}$

6. a) This equation is obtained by dividing $ax^2 + bx + c = 0$ by a and adding $b^2/4a^2$ to each side of the resulting equation.

b) This expression is the result of subtracting c/a from both sides of the equation obtained in part a).

c) We obtain this expression for x by taking square roots of both sides of the resulting equation in part b) and solving for x.

d) The results in parts a), b) and c) imply that if $ax^2 + bx + c = 0$, then $x = \frac{-b \pm \sqrt{b^2 - 4ac}}{2a}$.

7. a) axis of symmetry: $x = 0$ vertex: (0, 0)
h = ad^2 zeros: (0, 0)

b) axis of symmetry: $x = -b/2a$ vertex: $(-b/2a, -b^2/4a)$
h = ad^2 zeros: (0, 0) & $(-b/a, 0)$

c) axis of symmetry: $x = -b/2a$ vertex: $(-b/2a, c - b^2/4a)$
h = ad^2
The zeros are located on the x–axis, so their vertical distance, i.e. height h above the vertex is $h = b^2/4a - c$, or simply h = $(b^2 - 4ac)/4a$.
Since h = ad^2 then the distance d of the roots from the axis of symmetry is $\sqrt{h/a}$. That is, d $= \sqrt{b^2-4ac}\,/\,2a$.
Therefore the zeros are $(-b/2a - \text{d}, 0)$ and $(-b/2a + \text{d}, 0)$ or

$$\left(\frac{-b-\sqrt{b^2-4ac}}{2a}, 0\right) \text{ and } \left(\frac{-b+\sqrt{b^2-4ac}}{2a}, 0\right)$$

The roots of the equation $ax^2 + bx + c = 0$ are given by

$$x = \frac{-b \pm \sqrt{b^2 - 4ac}}{2a}$$

Answers to Exercises & Hints for Investigations

Exploration 5

1. a & b) c & d)

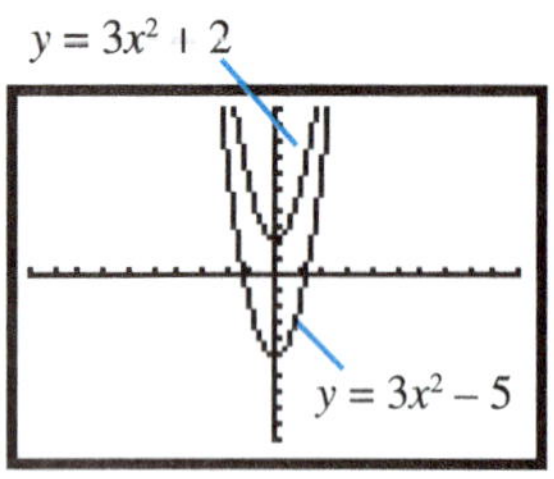

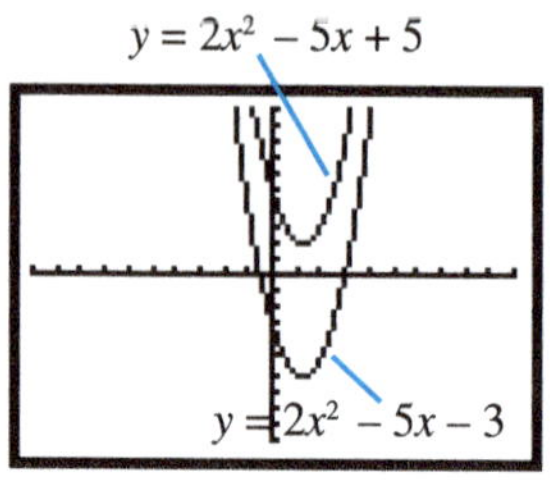

If we use the default setting $-10 \le x \le 10$; $-10 \le y \le 10$, and then trace without zooming in, we obtain these approximations:

a) $x = -1.276\ldots$ and $x = 1.276\ldots$

b) The curve does not cross the x-axis, so there is no root.

c) $x = -.425\ldots$ and $x = 2.978\ldots$

d) The curve does not cross the x-axis, so there is no root.

The [CALC] menu yielded the following roots:

a) $x = -1.29$ and $x = 1.29\ldots$

b) **ERR: SIGN CHNG** indicates there is no root.

c) $x = -0.5$ and $x = 3$

d) **ERR: SIGN CHNG** indicates there is no root.

2. The graph of $y = 3x^2 - 5$ intersects the x-axis in two points, so the corresponding equation has two real roots. However, the graph of $y = 3x^2 + 2$ lies entirely above the x-axis and therefore the corresponding equation has no real roots. Adding 7 to the function, $f(x) = 3x^2 - 5$ transforms it into the function $3x^2 + 2$ and this has the effect of translating its graph vertically upward until it no longer intersects the x-axis. Similarly the equation $2x^2 - 5x - 3 = 0$ has two real roots while the equation $2x^2 - 5x + 5 = 0$ has no real roots. This development suggests that it is always possible, by adding an appropriate constant, to transform a quadratic function with no real zeros into a quadratic function with two real zeros.

3. The graphs are shown below for the given window.

a)

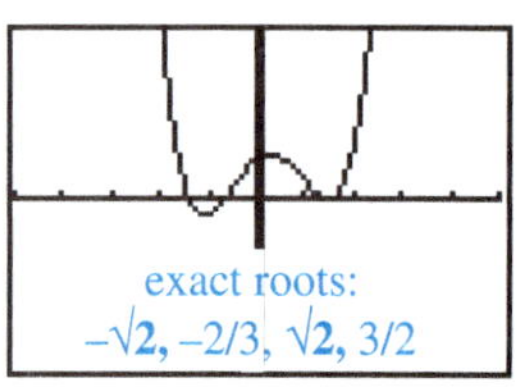

b)

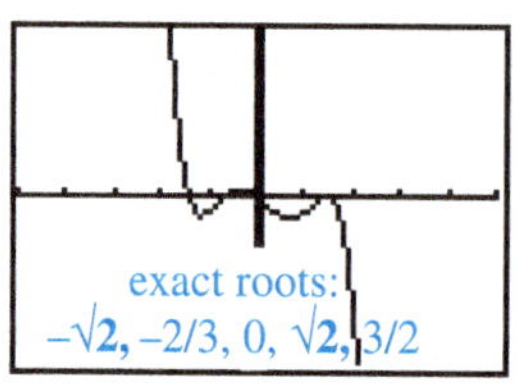

c)

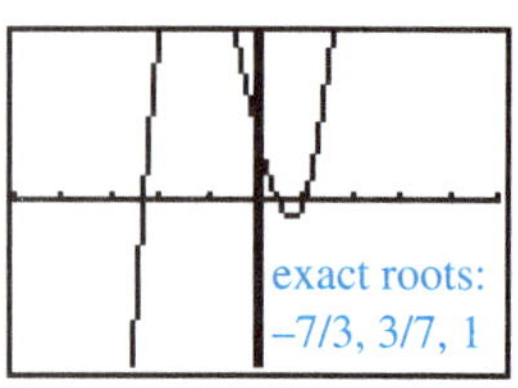

d)

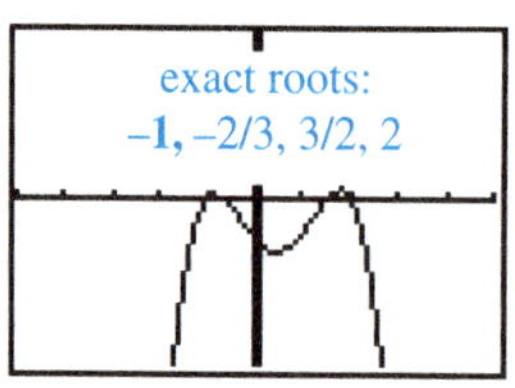

4. The *exact* roots are given in the displays above.

5. Using the [CALC] menu, we obtain the following roots

a) -2.17... -0.311... 1.481...

b) -0.938... 1.14... 2.79...

c) -0.493... 2.89... 5.603...

d) -2.82... -1.00... 2.00... 2.82...

Exploration 5 (cont'd)

6. The equation with roots a, b, c, is $(x - a)(x - b)(x - c) = 0$. Expanding and simplifying yields these equations.

a) $y = x^3 + x^2 - 4x - 4$

b) $y = x^4 - 4x^3 + x^2 + 8x - 6$

c) $y = x^4 - 5x^3 + 4x^2 + 6x$

7. a) A polynomial of degree n changes direction $n - 1$ times and therefore it would appear that it could cross the x-axis at most n times; so n is the maximum number of real roots.

b) If the polynomial is of degree n where n is even then we can slide the graph up or down by adding an appropriate constant so that the corresponding equation has no real roots. If n is odd, then the graph must cross the x-axis at least once so the corresponding equation has at least one real root.

c) no. (reason explained in part b)

Exploration 6

1. a) b)

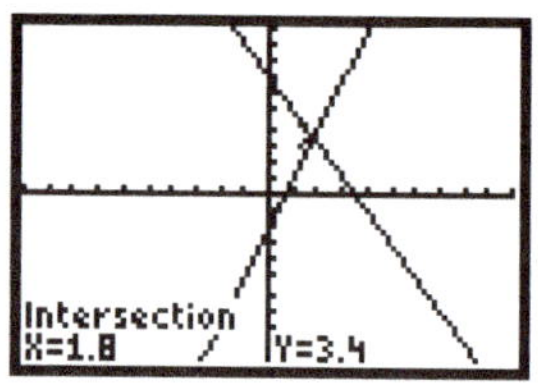

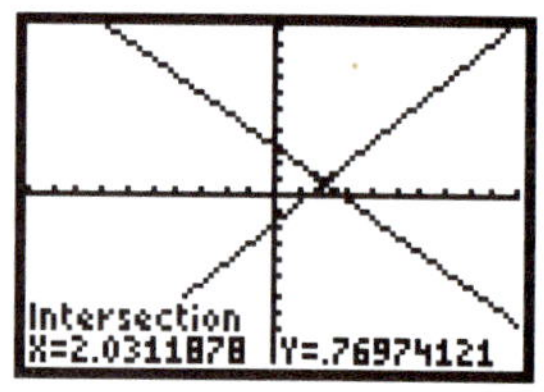

As shown in the displays, the roots obtained from using the [CALC] menu are respectively (1.8, 3.4) and (2.03…, 0.769…) The roots we obtain on solving the equations algebraically are respectively, (1.8, 3.4) and (2.03…,0.769…). All of the digits obtained using the [CALC] menu were correct.

2. a) The second equation is the same as the first equation with all coefficients multiplied by 5. Therefore the second equation is redundant and there are infinitely many solutions.

b) The graphs of the two equations are parallel lines and have no point in common. Therefore there is no solution to this system of equations.

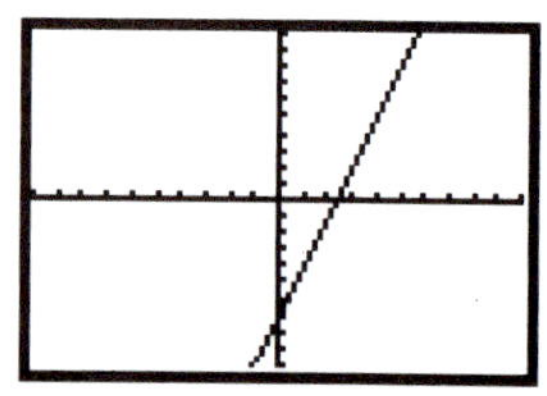

Display resulting from using 5:intersect.

3. The equations of the two lines are $y = 0.8x$ and $y = -0.2x + 6$. We solve this system to find that the point of intersection, (6, 4.8). i.e. 4.8 m above the ground.

4. Let x and y denote the time traveled by car (in hours) and the time traveled by ferry respectively.
Then, $x + y = 12.25$ and $80x + 16y = 740$. Solving, we obtain, $x = 8.5$, $y = 3.75$. That is, 8.5 hours were spent traveling by car.

Answers to Exercises & Hints for Investigations

Exploration 6 (cont'd)

5. A pair of linear equations has either no solution, one solution or an infinite number of solutions. These cases correspond respectively to the situations in which the two lines are parallel and are not coincident, the two lines are not parallel and therefore intersect in one point or the two lines are coincident.

6. The quadratic equation defines an ellipse and the linear equation defines a straight line. The line can intersect the ellipse in no points, one point (when it is tangent to the ellipse) or in two points. To determine which case applies, we graph the equations.

a) To graph the ellipse, we complete the square to obtain the equation:

$$\frac{\left(x-\frac{5}{2}\right)^2}{49}+\frac{y^2}{16}=1$$

which we express as the two equations

$$y=\pm\frac{4}{7}\sqrt{49-(x-5/2)^2}$$

The graphs of the equations are shown in the display. Using the intersection on the CALC menu, we find the following points of intersection.
$x = 1.996\ldots$, $y = 3.989\ldots$ and
$x = -0.534\ldots$, $y = -3.604\ldots$

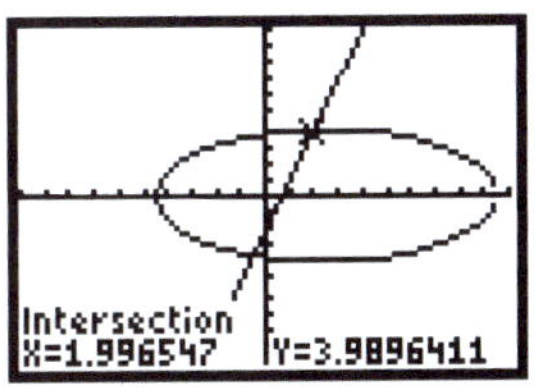

b) The graphs of the equations do not intersect, so there are no roots. If we seek an intersection using the CALC menu, we get an error message as shown.

7. The cubic curve intersects the ellipse in two points, so the system has two roots, shown in the displays below:

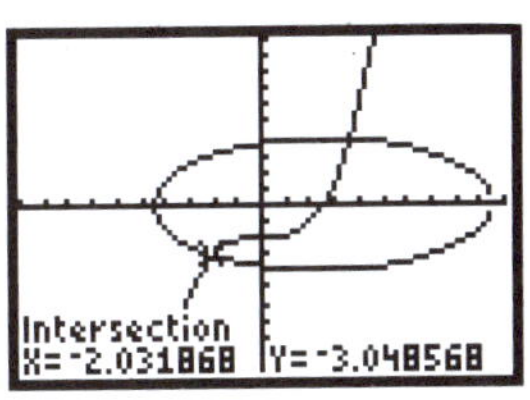

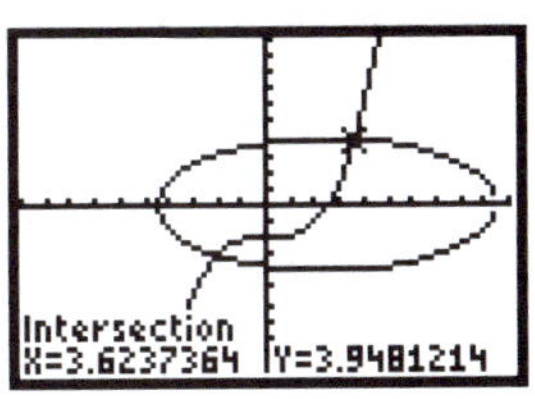

The roots are: $x = -2.031\ldots$ $y = -3.048\ldots$ $x = 3.623\ldots$ $y = 3.948\ldots$

Note: Remember when you use the [CALC] menu to find the intersection points, that the ellipse is composed of two curves in function mode. Therefore, in response to the prompts which ask for the first and second curves, ensure you use the "up" or "down" cursor key to select the two desired curves.

Exploration 7

1. Transcendental (i.e. non-algebraic) functions such as $\sin x$, $\log x$ and a^x cannot be represented by a finite number of algebraic expressions in x. Therefore, such functions satisfy Dirichlet's definition but not Euler's.

2. Euler's definition is contained in Dirichlet's definition, since any algebraic expression in x automatically defines a correspondence between x and $f(x)$; i.e. between x and y. Therefore there does not exist a function which satisfies Euler's but not Dirichlet's definition.

3. Since Cantor's definition requires a unique y associated with each x, a vertical line drawn through the graph of a function should intersect it in at most one point. Using this test, we discover that the graphs in "a" and "b" are not graphs of functions but the graph in "c" does correspond to a function.

4. The two points of intersection would have the same x-coordinate but different y-coordinates, violating the condition that there must be only one value of $f(x)$ for each x. Therefore, the graph would not be the graph of a function.

5. a) domain and range: the set of all real numbers
b) domain: the set of all real numbers range: $-9 \le y < \infty$
c) domain and range: the set of all real numbers
d) domain: $3 \le x < \infty$ range: $0 \le x < \infty$
e) domain: $x \le -4$ or $4 \le x$ range: $0 \le x < \infty$
f) domain and range: the set of all real numbers

6. a) $-25 \le y < \infty$ minimum at (-2, -25))
b) $-\infty < y \le -6.91$ maximum at (3.50, -6.75)
c) the set of all real numbers no extrema
d) $-3.81 \le y < \infty$ mimimum at (1.31..., -3.845)

Note: For the quintic polynomial in part c) we get a good view of the behavior of the function by setting the range variables to:
$-10 \le x \le 10$ $-1000 \le y \le 1000$

7. a) vertex: (1, -36) axis of symmetry: $x = 1$
To get close to these values you may *zoom in* on the vertex using the keying sequence: [ZOOM] 2 [ENTER], or you can find the minimum using the [CALC] menu.

b) vertex: (-9.5, 107.25) axis of symmetry: $x = -9.5$
(See note in part a.)

Hints for the Investigations:

To investigate the relationship between the equation of the axis of symmetry of the parabola defined by the equation: $ax^2 + bx + c = 0$ and the coefficients a and b, make a table which lists for various values of a and b, the value of $\frac{b}{a}$ and the equation of the axis of symmetry of the corresponding parabola.

Answers to Exercises & Hints for Investigations

Exploration 8

1. For the function $f(x) = 0.9x - 0.0014x^2$, the domain is the set of real numbers and the range is $-\infty < y \leq 144.64$. However, the trajectory of the baseball is represented by the function $f(x)$ only over the domain $0 \leq x \leq 321.43$ and over this domain its range is $0 \leq y \leq 144.64$.

2. a) a parabola b) 321.43 feet
   c) the slope is $144.64 / 321.43 = 0.450$.
   d) When the baseball reaches its highest point, its rate of change of height must be zero, for otherwise it would travel higher. That is, the slope of the tangent to the trajectory is 0.
   e) This mathematical model neglects the effect of wind resistance which is greatest when the ball is moving fastest. This distorts somewhat the symmetry of the trajectory. The model also assigns a height of 0 to the ball at the initial moment of impact, while we know that the ball is actually 3 or 4 feet above the ground. In spite of these limitations, the model gives us a reasonable approximation to the true path of the ball.

3. a) A function $f(x)$ has a local minimum at $x = a$ if and only if $f(x) \geq f(a)$ for all values of x close to a.
   b) No; the curve defined by $y = x^3 + 8$ is monotonically (steadily) increasing, so as you trace from left to right along the curve, the x and y values increase together.

4. Graph of $f(x)$ in the window:
   $-10 \leq x \leq 10$; $-200 \leq y \leq 100$

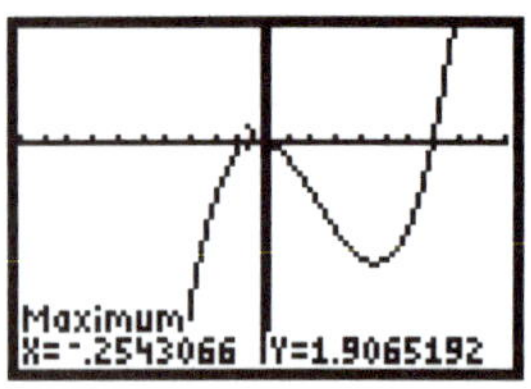

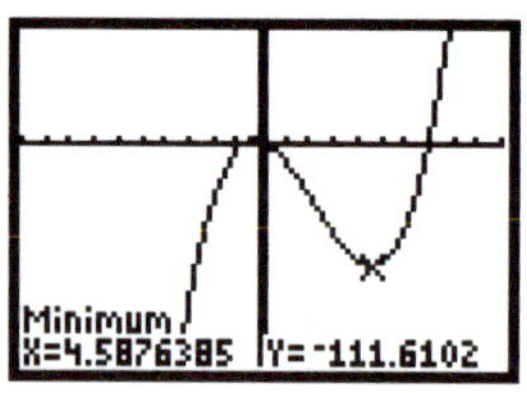

Using the [CALC] menu and selecting in turn the maximum and the minimum, we obtain the displays shown above. As the displays show, the maximum of 1.906... occurs when $x = -0.254...$ and the minimum of $-111.610...$occurs when $x = 4.587...$

5. a) Using the [CALC] menu, we obtain the display shown on the right. This quadratic function has a minimum at (1.166..., 0.9166...)
   b) Using the [CALC] menu, we obtain the displays shown below. This function has a minimum at (1.00..., –8.00...). It also has a minimum at (–2.00..., –8.00...) and maximum at (–0.50, –2.94...).
   c) This function has a minimum at (0.50, –6.687...). It has no other extrema.

a)

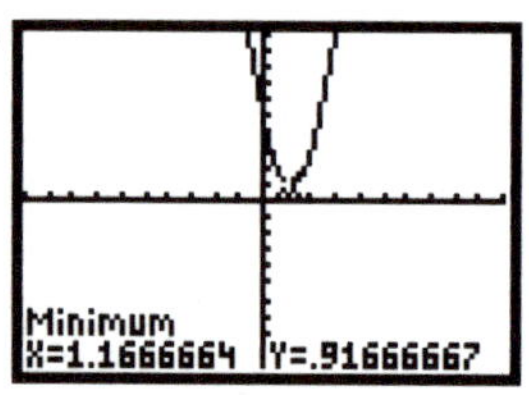

b)

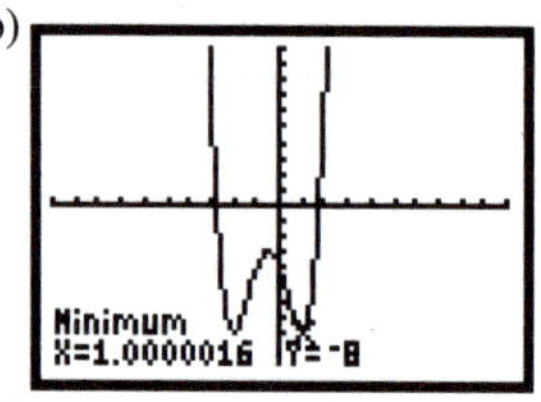

6. a) none; graph has fixed slope and cannot change direction.
   b) one; graph is a parabola which opens upward or downward.
   c) two; graph can cross x-axis 3 times; it can change direction twice
   d) By induction one might conjecture $n - 1$. (It can be proved.)

Exploration 9

1. Any number which can be expressed as a quotient of two integers is called a *rational number.* Similarly, any function which can be expressed as a quotient of two polynomials is called a *rational function.*

2. Yes. A polynomial is a special rational function of the form $\frac{f(x)}{g(x)}$ where g(x) is an integer or a factor of $f(x)$.

3. A singularity of a function is a point at which the function becomes infinite or undefined. The singularities of a rational function are those points at which the denominator of the function is zero. Therefore, to find the singularities of a rational function, we must find the values of the variable for which the denominator is 0.

4. If $f_1(x)$ and $f_2(x)$ are respectively, $\frac{g_1(x)}{h_1(x)}$ and $\frac{g_2(x)}{h_2(x)}$

   then $f_1(x) \pm f_2(x) = \frac{g_1(x)h_2(x) \pm g_2(x)h_1(x)}{h_1(x)h_2(x)}$ which are rational functions.

   b) The product and quotient of $f_1(x)$ and $f_2(x)$ are respectively, $\frac{g_1(x)\ g_2(x)}{h_1(x)\ h_2(x)}$ and $\frac{g_1(x)\ h_2(x)}{g_2(x)\ h_1(x)}$ which are rational functions.

5. The singularities are the points at which the denominator is zero.
   a) at $x = 0$ b) at $x = 2$
   Note: To remove the singularity at $x = 3$, we divide the numerator and denominator by the common factor $x - 3$.
   c) Since the discriminant of $x^2 - 3x + 7$ is negative, this polynomial has no real zeros. Therefore $f(x)$ has no singularities.

As $x \to \infty$ $\frac{1}{x} \to 0$; $\frac{x-3}{x^2 - x + 6} \to 0$; $\frac{x^3}{x^2 - 3x + 7} \to \infty$

As $x \to -\infty$ $\frac{1}{x} \to 0$; $\frac{x-3}{x^2 - x + 6} \to 0$; $\frac{x^3}{x^2 - 3x + 7} \to -\infty$

6.
a)

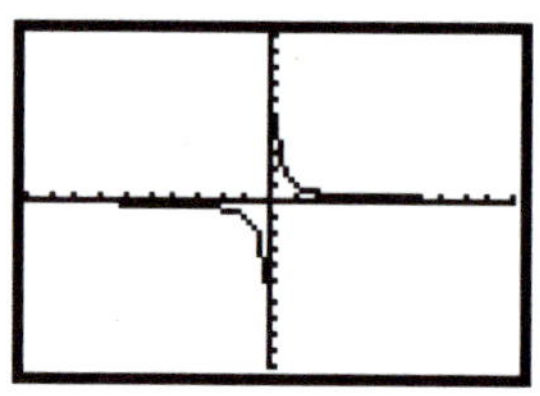

b)

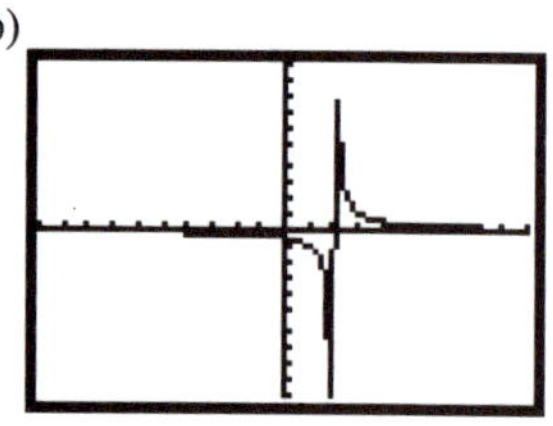

c)

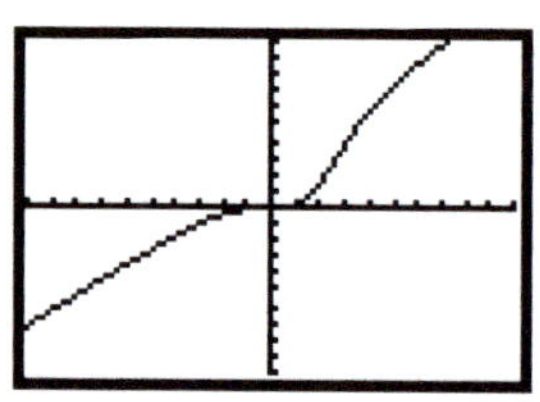

We see in example b) above, that although the denominator is zero at $x = 3$, so also is the numerator. By cancelling the common factor $x - 3$, we redefine the function at $x = 3$ to have the value $f(3) = 1$.

Answers to Exercises & Hints for Investigations

Exploration 9 (cont'd)

7. $f(x) = \dfrac{1}{(x-4)(x-7)}$ or $\dfrac{1}{x^2 - 11x + 28}$

8. a) b)

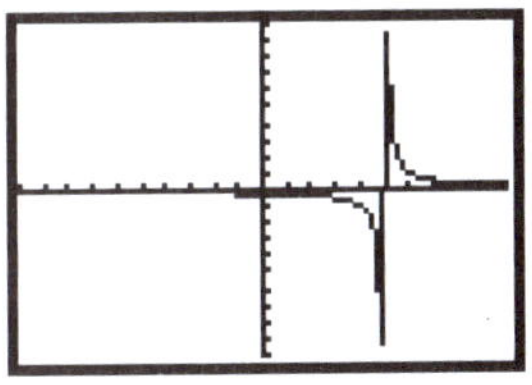

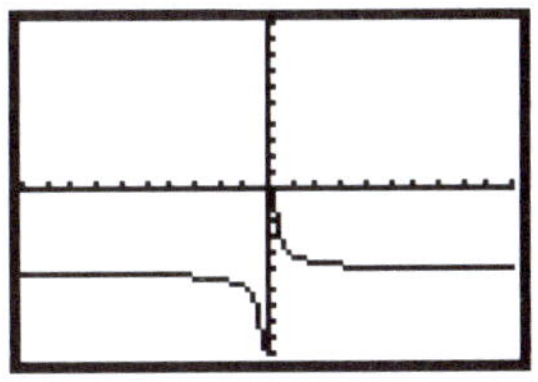

The keying sequences are as follows:

a) [Y=] [1] [÷] [(] [X,T,θ,n] [−] [5] [)] [ENTER] [GRAPH]

b) [Y=] [1] [÷] [X,T,θ,n] [−] [5] [ENTER] [GRAPH]

The difference in the keying sequences is that in part a) we used brackets to group the $x - 5$, while in part (b) we did not. Therefore the singularities are respectively at $x = 5$ and $x = 0$.

Both functions approach 0 as $x \to -\infty$ and as $x \to \infty$.

9. For $\dfrac{1}{x-5}$ [Y=] [(] [X,T,θ,n] [−] [5] [)] [x^{-1}] [ENTER] [GRAPH]

For $\dfrac{1}{x} - 5$: [Y=] [X,T,θ,n] [x^{-1}] [−] [5] [ENTER] [GRAPH]

Using the [x^{-1}] key saves us one stroke when the numerator is 1.

10. a) b)

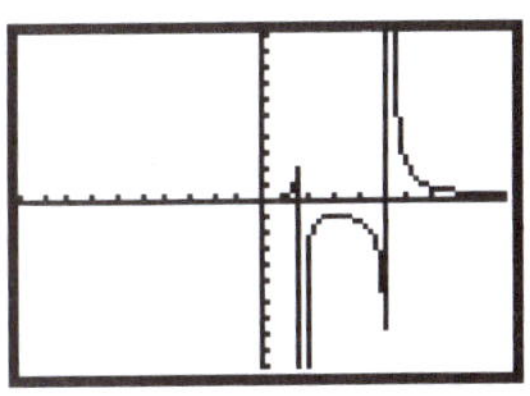

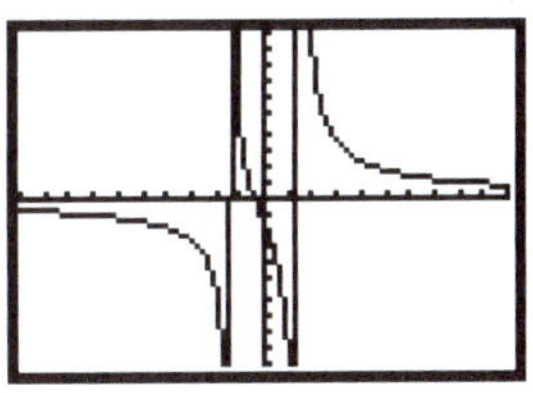

a) The graph above shows that there are singularities near $x = 1.7$ and near $x = 5.3$. To *zoom in* on the singularity near $x = 1.7$, we press [TRACE] and cursor to the right until the sign of the y-coordinate jumps from a positive value to a negative value. This occurs when $x = 1.7021277$. To obtain a closer approximation to the singularity, we then zoom in with the sequence: [ZOOM] 2 [ENTER]. Once the curve has been re-drawn, we press [TRACE] and move the cursor left once until the y-coordinate changes from negative to positive. This occurs when $x = 1.6489362$. We repeat this entire *zoom in* procedure 3 more times until we obtain $x = 1.6954787$. If we press the cursor key once more, we jump to the point with $x = 1.6988032$ and $y = -298.5564$. Therefore, the singularity lies in the interval $1.695 \le x \le 1.699$. That is; $x \approx 1.70$ correct to 2 decimal places. To obtain a closer approximation to the position of the singularity near $x = 5.3$, we proceed as above, using [ZOOM] 2 [ENTER] followed by [TRACE]. We repeat this process three times to determine that the singularity is at $x \approx 5.30$ (correct to 2 decimal places).

Investigation: Try finding the singularities of this function by finding the zeros of its reciprocal.

Exploration 9 (cont'd)

Note: When using the [ZOOM] 2 [ENTER] to obtain a closer approximation to the location of a singularity, it is important to wait until the curve has been re-drawn and the busy signal in the upper right corner of the screen has vanished before pressing [TRACE]. This will ensure that the cursor travels along the curve when you press the left and right cursor keys.

Domain of $x(9 - 7x + x^2)^{-1}$ is: All real x except $x \approx 1.70$ & 5.30.
Range of $x(9 - 7x + x^2)^{-1}$ is the set of all real numbers.

We observe from the graph that the curve approaches 0 asymptotically as x approaches $\pm\infty$ and there seems to be a local maximum near $x = 3$. To find the coordinates of this local maximum, we press [TRACE] and move the cursor along the curve to $x = 2.9787234$, $y = -1.000152$. Then we press [2nd] [CALC] 4. In response to the prompt for a lower bound, we move the cursor left and press [ENTER]. In response to the prompt for an upper bound, we move the cursor right of the top of the curve and press [ENTER]. In response to the prompt for a guess, we move the cursor to the apparent maximum and press [ENTER]. We obtain the maximum point, **X = 3.0000005, Y= -1.** i. e. a local maximum occurs at (3, -1).

b) To find the locations of the singularities, correct to two decimal places, we could proceed as in part a) (See note above.) The graph indicates there are singularities near $x = -1.4$ and 1.4. By tracing along the curve we discover the singularity is right of the point (1.2765957, –32.2335). Alternatively, we observe that the denominator is zero when $x^2 - 2 = 0$, i.e. when $x = \pm\sqrt{2}$, so the singularities occur at $x = \pm 1.414\ldots$

Domain of $(7x + 3)(x^2 - 2)^{-1}$ is: All real x except $x \approx \pm 1.414\ldots$
Range of $(7x + 3)(x^2 - 2)^{-1}$ is the set of all real numbers.
The graph suggests there are no local maxima or minima.

c) When we graph this function with the default window settings, we obtain the graph shown below left. There appear to be singularities near $x = -3$ and $x = 4$. We can proceed to use [TRACE] followed by [ZOOM] 2 [ENTER] as in parts a) and b) above and we confirm that the singularities are in fact, at $x = -3$ and $x = 4$. Therefore the domain of $x^3/(x^2 - x - 12)$ is the set of all real x except $x = -3$ and $x = 4$. The range is the set of all real numbers because y assumes every real value. (i.e. the projection of the graph on the y-axis is the y-axis.)

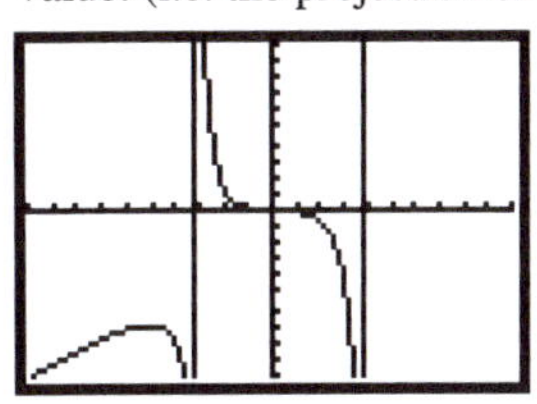

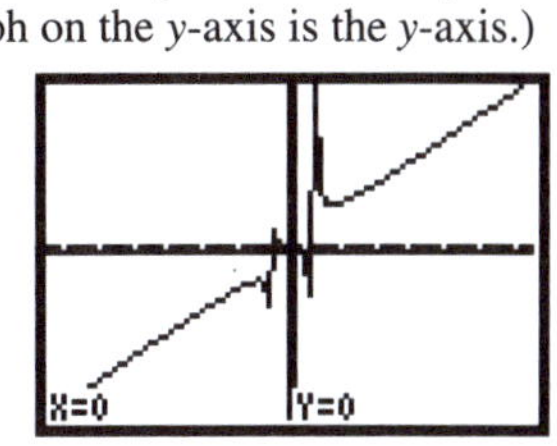

From the graph in the figure (above left), we can see that the function has a local maximum near $x = -5$. None of the graph is visible to the right of the singularity at $x = 4$. Therefore we zoom out and get the graph shown on the right. We see that there is a local minimum near $x = 7$. Proceeding as in parts a) and b) we find a local maximum at (–5.082763, –6.941315) and a local minimum at (7.082761, 11.431).

Answers to Exercises & Hints for Investigations

Exploration 10

1. An optimization problem is any problem in which an extremum of a function is sought subject to a constraint on one or more of the variables.

a) $S = 2\pi r^2 + 2\pi rh$ b) $\pi r^2 h = 300$

2. Let x denote the length. Then the width is $1.8 - x$. We must maximize Area $= x(1.8 - x)$. We graph $y = -x^2 + 1.8x$. and obtain the parabola shown in the display. We use the [CALC] menu and select 4 (for maximum). After responding to the prompts for lower and upper bounds, we enter a guess. The display yields **X = .90000102; Y = .81** as shown. Therefore the maximum area is enclosed when the rectangle has length of 0.9 m and width 0.9 m. That is, the rectangle is a square.

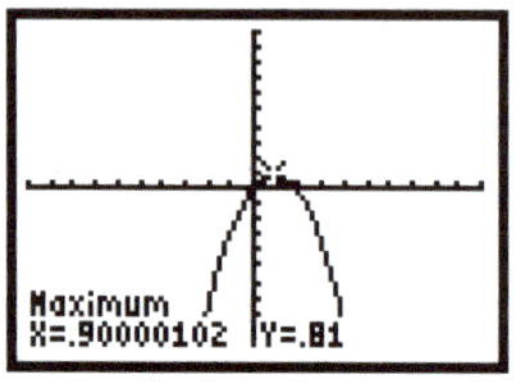

3. The distance of the point (x, y) from the origin is $\sqrt{x^2+y^2}$. We must minimize this function given the constraint $y = 3x + 2$. That is, we must minimize $f(x) = \sqrt{x^2 + (3x+2)^2}$. We graph this equation and select 3 (for minimum) from the [CALC] menu. After responding to the prompts for lower and upper bounds, we enter a guess. We obtain the display shown here. This indicates that the point with x-coordinate -0.6 is the point on the line $y = 3x + 2$ which is closest to the origin. This is the point (-0.60, 0.2). Its distance from the origin is 0.63 units correct to 2 decimal places.

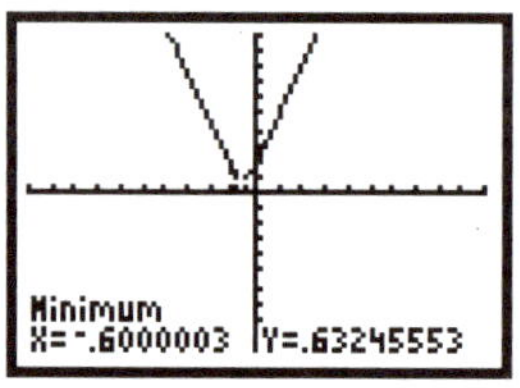

4. We define $Y_1 = x^2 - 4x + 3$ and $Y_2 = \sqrt{x^2 + Y_1^{\,2}}$
To minimize Y_2 , we graph it and select 3 (for minimum) on the [CALC] menu. After responding to the prompts for lower and upper bounds, we enter a guess. We obtain the display shown here. The display indicates that the point on the parabola with x-coordinate 0.83… is the point closest to the origin. It's distance to two decimal places is 0.91 units.

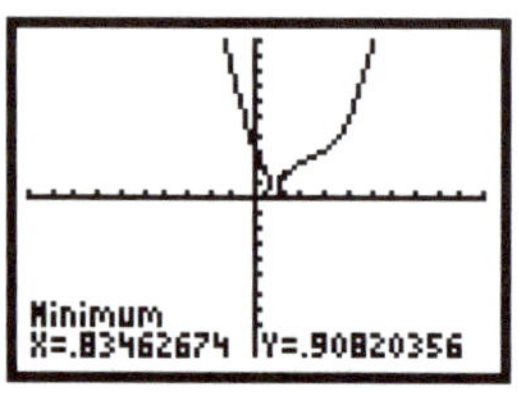

5. Since the lines are parallel, all points on the line $y = 2.7x - 13$ are equidistant from the line $y = 2.7x + 16$. Therefore, it suffices to choose any point on the line $y = 2.7x - 13$, such as (0, –13) and determine its distance from the line $y = 2.7x + 16$. That is, we must minimize the function $\sqrt{x^2 + (y + 13)^2}$ subject to the constraint $y = 2.7x + 16$. That is, we must minimize the function: $f(x) = \sqrt{x^2 + (2.7x + 29)^2}$. When we graph this function, we must zoom out to have the graph appear on the screen. To minimize $f(x)$, we graph it and select 3 (for minimum) on the [CALC] menu. After responding to the prompts for lower and upper bounds, we enter a guess. We obtain the display shown here. The display indicates that the point on the line $y = 2.7x + 16$ closest to (0, -13) has x-coordinate, -9.45. The distance between the two points and hence the two lines is 10.07 units (to 2 decimal places).

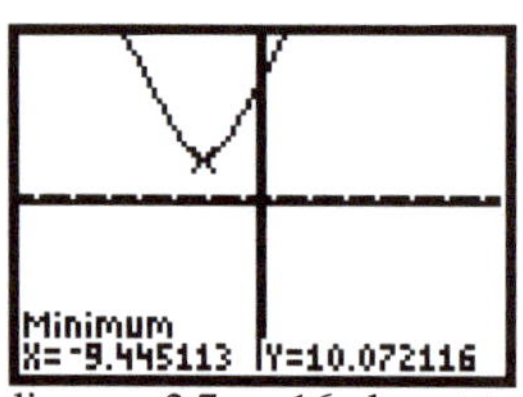

Exploration 10 (cont'd)

6. The line perpendicular to the line $y = 2.7x + 16$ and passing through the point (0, -13) has equation $y = -\frac{x}{2.7} - 13$. We graph these two equations. After zooming out once, we obtain the graphs shown in the display. We select 5 (intersect) from the [CALC] menu. After responding to the prompts for "first curve", "second curve" and "guess", we obtain the point of intersection, **X = -9.445115; Y= -9.501809.** That is, the point, (-9.445, -9.502) is the point of intersection (to 3 decimal places). The distance between the lines defined by $y = 2.7x - 13$ and $y = 2.7x + 16$ is therefore the distance between the points (0, -13) and (–9.445, –9.502). This distance is given by $\sqrt{9.445^2 + (13 - 9.502)^2}$ or 10.07 (to 2 decimal places).

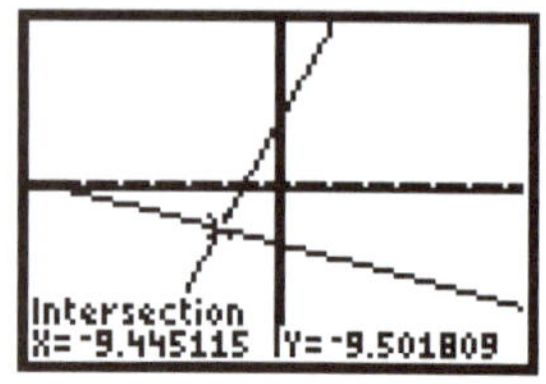

7. The position of the luxury liner is given by $(0, Y_1)$ where $Y_1 = -42x$ and x denotes the number of hours past 7:00 a.m. The position of the cargo ship is given by $(Y_2, 0)$ where $Y_2 = 80 - 48x$. We must minimize $\sqrt{Y_1^{\,2} + Y_2^{\,2}}$. We define Y_1 and Y_2 . Using the [Y-VARS] menu, we define $Y_3 = \sqrt{Y_1^{\,2} + Y_2^{\,2}}$. If we attempt to graph Y_3 using the default settings, we find that the graph is not visible. Since $x > 0$ and $Y_3 > 0$, and since the cargo ship will take less than 2 hours to reach the origin, we set $X_{min} = 0$, $X_{max} = 2$; $Y_{min} = 0$, $Y_{max} = 100$

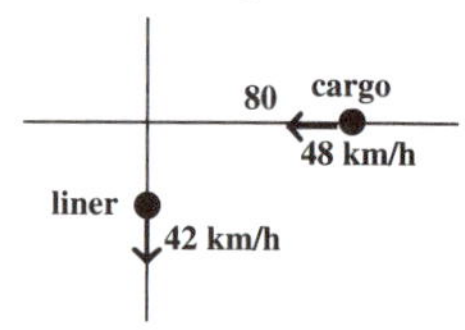

Graphing Y_3 in this window yields the display on the right. We select 3 (minimum) on the [CALC] menu. After entering a lower bound, an upper bound and an estimate we obtain the minimum point, shown.

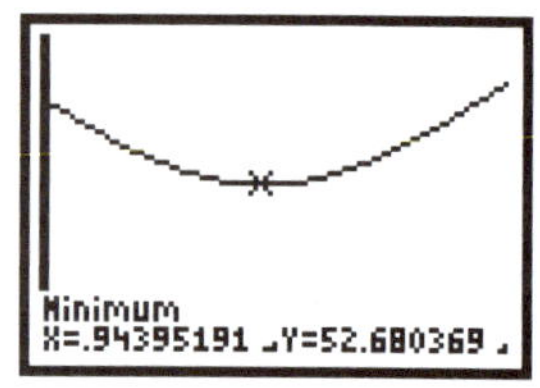

That is, the ships reach their closest proximity at 0.94 hours or 56.4 minutes after 7:00 a.m. At that time they are 52.68 km apart.

8. We define the cost and sales functions as follows:
$Y_1 = 0.35x^2 + 23x + 20$; $Y_2 = 80x$
The profit function to be maximized is: $Y_3 = Y_2 - Y_1$.
We graph Y_3 for the range:
$0 \le x \le 200$; $-5000 \le y \le 5000$
We then select 4 (maximum) on the [CALC] menu. Upon responding to the prompts for a lower bound, upper bound and an estimate, we obtain the display on the right. This indicates that the optimum daily production is about 81 grummets which yields a profit of about \$2300.

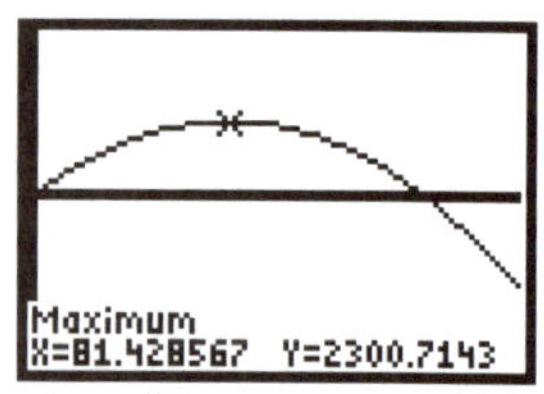

9. If x denotes the width of Ms. Chiu's pool, then the area is given by $Y_1 = (35-2x)x$. We set the range values to: $0 \le x \le 18$; $0 \le y \le 500$ and graph Y_1. Proceeding as in exercise 8, we obtain the display on the right.

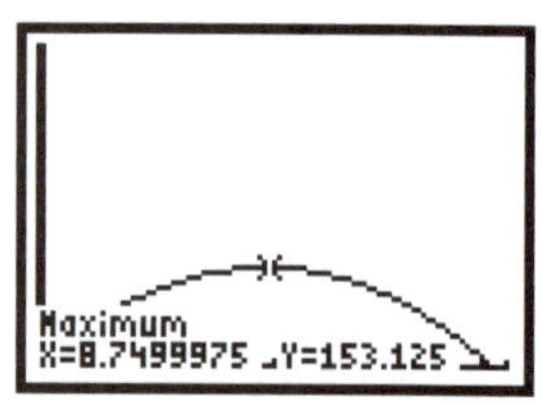

The optimal dimensions are: length 17.5 m and width 8.75 m.

Answers to Exercises & Hints for Investigations

Exploration 11

1. a) $R^2 = V^2 + (W+T)^2$
$= V^2 + W^2 + T^2 + 2TW$
$= S^2 + T^2 + 2TW$

b) $W = S\cos(180 - \theta) = -S\cos\theta$

c) $R^2 = S^2 + T^2 + 2T(-S\cos\theta)$
$= S^2 + T^2 - 2ST\cos\theta$

2. a) In the special case when $\theta = 90°$, $\cos\theta = 0$ and the cosine law reduces to the Pythagorean Theorem.

b) Setting $R = 80$ and $S = T = 60$ in the cosine law yields:

$$80^2 = 60^2 + 60^2 - 2(60)(60)\cos\theta.$$

Solving for $\cos\theta$, we obtain $\cos\theta = 1/9$. Hence $\theta = 83.62°$.

3. The distance from a corner of the square base to the center of the base is half the distance to the opposite corner. That is, the distance is half of $230\sqrt{2}$ or $115\sqrt{2}$ m. The corner, the center of the base and the tip of the pyramid form a right triangle with height 147 m and base $115\sqrt{2}$ m. If θ is the angle of elevation, then $\tan\theta = 147/115\sqrt{2}$. That is, $\theta = 42.11°$.

4. The diagram shows the first tee shot located 212 yards from the tee and 16 yards, i.e., 48 feet, from the hole. The second ball is located 203 yards from the tee and x yards from the hole. Applying the cosine law, we obtain:

$(x + 16)^2 = 212^2 + 203^2 - (212)(203)\cos 8°$.

from which we find $x = 14.31$ yards. Therefore the second ball is 1.69 yards or about 5 feet closer. The second golfer won the car.

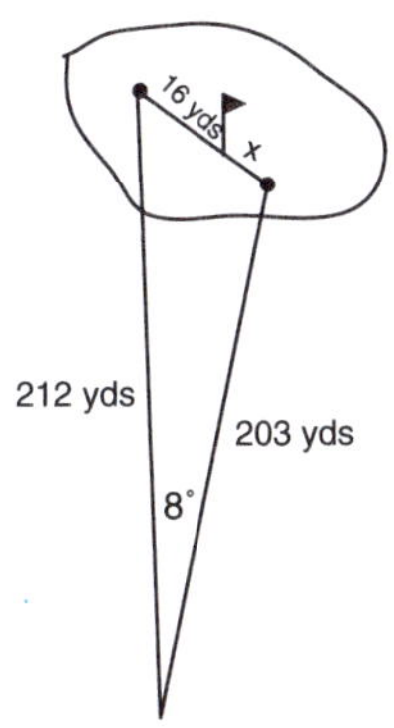

5. After 6 minutes, one cyclist is 5.5 miles from the intersection and the other cyclist is 4.8 miles from the intersection. Applying the cosine law, we determine that the distance between the two cyclists is 3.3 miles.

6. When we run ALLSIDES and enter the values we obtain this display. The largest sides are opposite the largest angles:

side	angle
45.2	65.35
44.6	63.74
38.6	50.91

```
SHORTEST SIDE38.
6
THIRD SIDE44.6
ANGLES ARE
           65.35
           50.91
           63.74
```

7. The angles are the same as in exercise 6 because the two triangles are similar; that is, each triangle is an enlarged or reduced version of the other.

8. When we run ALLSIDES, we discover that the angles in degrees are: 90, 18.92 and 71.08. That is, these sides form a right triangle, so 12, 35 and 37 is a Pythagorean triple.

9. When we run SAS, we find that the distance from tee to green over the pond is 187.72 yards. You might consider a 5 iron if there is no headwind, or a 3 iron if you can't swim.

10. a) $\cos\theta = 6370/6370.553$ and so $\theta = 0.01317$ radians.

b) When θ is in radians, $s = 6370 \times \theta$ or 83.93 km.

c) About 84 km.

Exploration 12

1. When we run SSA, we discover that there are two triangles which satisfy these criteria. (called the ambiguous case). The two triangles are:

Triangle 1:	Sides:	2.76	3.32	2.70
	Angles:	53.38	74.90	51.72
Triangle 2:	Sides:	2.76	3.32	1.26
	Angles:	53.38	105.10	21.52

2. a) AD = bsin C and AD = csinB

b) bsinC = csinB

c) Draw a diagram and verify that the same relationships hold.

d) Division by sinBsinC yields b/sinB = c/sinC.
Similarly, asinB = bsinA, so we divide by the product of the sines to complete the proof of the sine law.

3. When we run program AAS, the display shows that the measures of the angles are: 64°, 38° and 78°. The measures of the sides are: 21.6, 14.8 and 23.5.

4. Upon running SAS, we find that the missing side is 78.62 km, so the distance between the satellites is 78.62 km.

5. The angle opposite the side of length 300 m has measure of 11.8°. We now apply AAS and enter the values of the side length, and the two angles to obtain 1113.86 m the length of the longest side. The height of the mountain is 1113.86sin37.6 or 679.6 m.

6. Program TRIANGLE, below left, takes as input any 3 measures of the sides and angles of a triangle. It then branches to the appropriate subroutine (programs ALLSIDES, AAS, SAS, and SSA) displaying all the angles and sides of the triangles which fit the input data. Branching is based upon the number of known angles and contained angles. TRIANGLE yields these displays.

```
PROGRAM:TRIANGLE
:ClrHome
:Disp "HOW MANY"

:Input "KNOWN AN
GLES",W
:If W=0
:Then
:prgmALLSIDES
:Else
:If W=2
:Then
:prgmAAS
:Else
:Disp "HOW MANY"
:Disp "CONTAINED
"
:Input "ANGLES K
NOWN?",Z
:If Z=1
:Then
:prgmSAS
:Else
:End
:prgmSSA
:End
:End
:
```

Program TRIANGLE shows how a series of subroutines can be combined to create a powerful and comprehensive program.

a)

```
OPPOSITE ANGLE?8
0
OTHER ANGLE?60
ANGLES ARE
              80
              60
              40
SIDES ARE
            17.6
     11.48758415
     15.47718025
            Done
```

b)

```
OTHER SIDE?26.5
TWO TRIANGLES
PRESS ENTER
ANGLES
              40
     49.80485746
     90.19514254
SIDES
            22.3
            26.5
     34.69244012
ANGLES
              40
     130.1951425
     9.80485746
SIDES
            22.3
            26.5
     5.907915364
            Done
```

Answers to Exercises & Hints for Investigations

Exploration 13

1. a) One *radian* is the measure of the angle at the center of a circle which is subtended by an arc whose length is equal to the radius of the circle. Since the circumference of a circle has length $2\pi R$, a full turn (i.e. the angle at the center of the circle subtended by the circumference) has a measure of 2π radians.

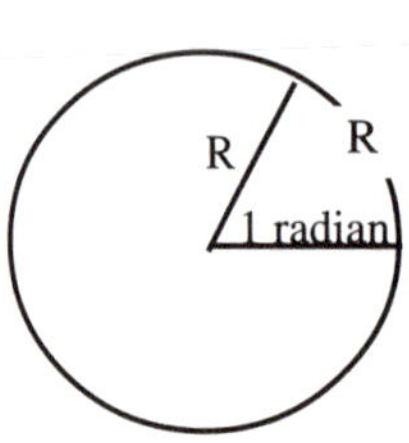

b) From part a), it follows that 2π radians $= 360°$, or π radians $= 180°$. Therefore to convert degrees to radians, we multiply the number of degrees by $\frac{\pi}{180}$

To convert radians to degrees multiply the number of radians by $\frac{180}{\pi}$.

2. Use the formula in 1b) to convert the equation to:

$$y = \frac{35}{3} + \frac{7}{3}\sin\frac{2\pi(x-80)}{365}$$

3. The new window settings are:

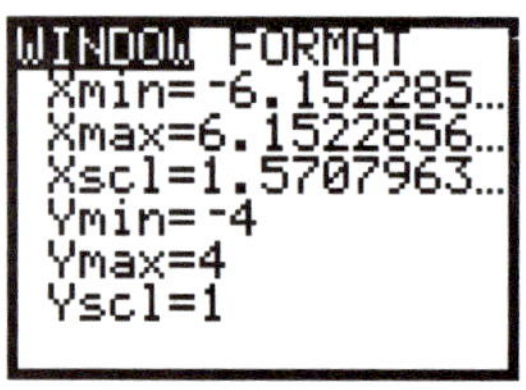

4. The display below shows the graphs of all three functions for the ZTrig (Zoom 7) window settings. The function defined by $y = 3\sin x$ has the highest amplitude and the function defined by $y = \sin x$ has the lowest.

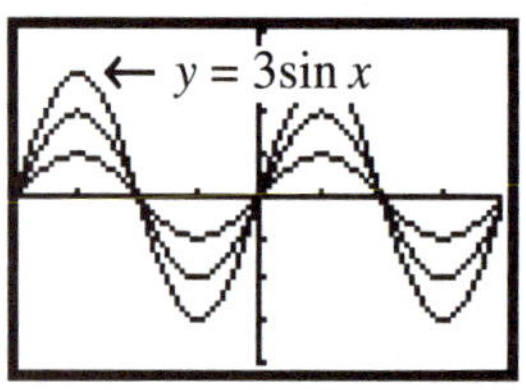

As A is increased by a factor of k, the amplitude of the graph is increased by a factor of k.

5. We define the functions shown in the display and graph them as in *worked example* 2 using the window settings as shown below.

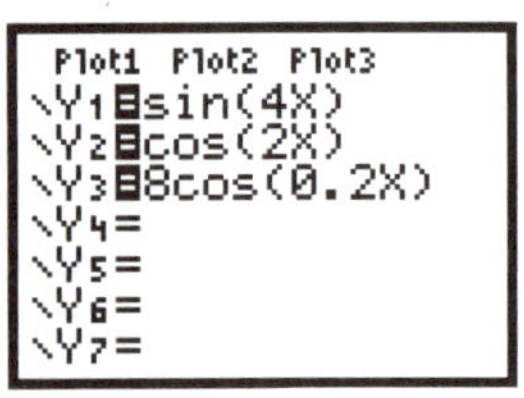

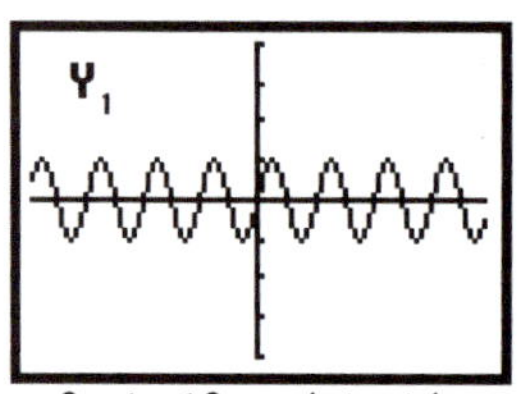

$-2\pi \le x \le 2\pi$; $-4 \le y \le 4$

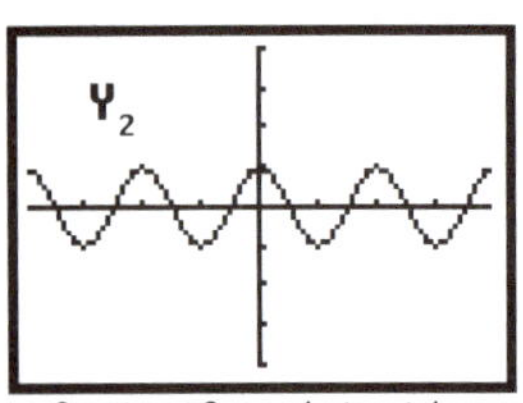

$-2\pi \le x \le 2\pi$; $-4 \le y \le 4$

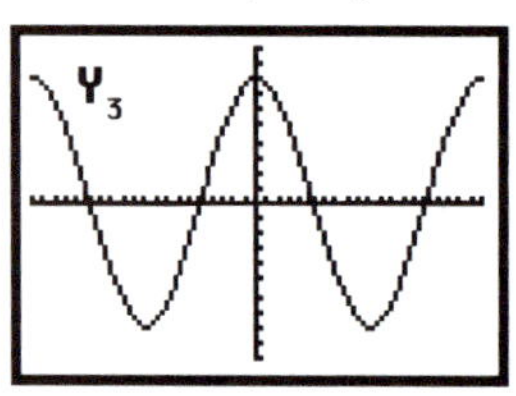

$-10\pi \le x \le 10\pi$; $-10 \le y \le 10$

We observe that for $0 < x < 2\pi$, the functions, Y_1, Y_2, and Y_3 have periods of $\pi/2$, π and 10π. It would appear that the period of $y = \sin Bx$ and $y = \cos Bx$ may be $2\pi/B$. The analytical proof is left as an exercise.

Exploration 13 (cont'd)

6. We define $Y_1 = \sin x$, $Y_2 = \sin(x + \pi/3)$ and $Y_3 = \sin(x + \pi/2)$. Using the ZTrig settings (zoom 7), we obtain the graphs shown in the display below.

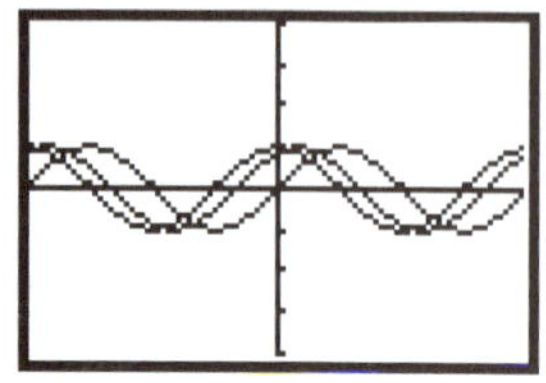

If we graph each function separately, we can observe that as C ($C > 0$) is increased, the graph shifts to the left (translates) by the amount that C is increased. If $C < 0$ and C is further decreased, the graph is translated to the right by the amount that C is decreased.

Exploration 14

1. We consider the graphs of $f(x) = b^{Ax}$ for $0 < b \le 1$ and for $b > 1$. (Since b^{Ax} is not a real-valued function for $b < 0$, we need only consider positive values of b.) To graph $f(x) = b^{Ax}$ for $A = 2$ and $b = 0.5$, 1 and 2, we define $L_1 = \{0.5, 1, 2\}$ and define $Y_1 = L_1$ ^$(2x)$. Graphing Y_1, we obtain the display shown here. From these graphs, it is clear that:

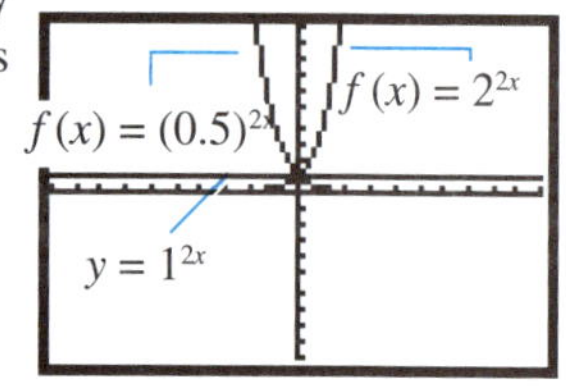

a) as $x \to \infty$
slope $b^{Ax} \to \begin{cases} 0 & \text{if } 0 < b \le 1 \\ \infty & \text{if } b > 1 \end{cases}$

b) as $x \to -\infty$
slope $b^{Ax} \to \begin{cases} 0 & \text{if } b \ge 1 \\ -\infty & \text{if } 0 < b < 1 \end{cases}$

2. We could proceed as in exercise 1, by graphing b^{Ax} for $A = 2$ and the same values of b. However, we need only observe that

$$\begin{aligned}(0.5)^{-2x} &= (1/2)^{-2x} \\ &= 2^{2x}\end{aligned}$$

so the graph of $f(x) = (0.5)^{-2x}$ is the same as the graph of $f(x) = 2^{2x}$. Similarly, we can show that the graph of $f(x) = 2^{-2x}$ is the same as the graph of $f(x) = (0.5)^{2x}$. Therefore, using the results of exercise 1, we have,

a) as $x \to \infty$
slope $b^{Ax} \to \begin{cases} \infty & \text{if } 0 < b < 1 \\ 0 & \text{if } b \ge 1 \end{cases}$

b) as $x \to -\infty$
slope $b^{Ax} \to \begin{cases} 0 & \text{if } 0 < b \le 1 \\ -\infty & \text{if } b > 1 \end{cases}$

3. In theory, it would never decay entirely, because half its radioactive portion decays every 500 years; therefore, at the end of every 500 year period there is always 50% remaining which has not yet decayed.

Answers to Exercises & Hints for Investigations

Exploration 14 (cont'd)

4. To graph these six functions, define $L_1 = \{-1, -0.5, -0.25, 0.25, 0.5, 1\}$. Then graph $Y_1 = 8\text{^}(L_1x)$. We obtain the display on the right.

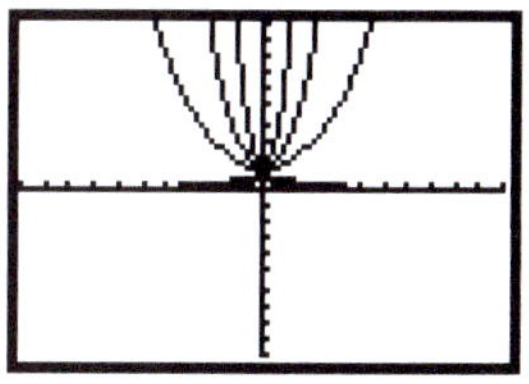

a) The graph of $f(x) = 8^{Ax}$ becomes steeper and moves closer to the y-axis as $|A|$ increases.

Recall: We can move vertically from one curve to another by pressing the [▲] and the [▼] keys.

b) Since $(8^{0.5x})(8^{0.5x}) = 8^x$, we observe that 8^x is the square of $8^{0.5x}$ for any x. Therefore, any point on the graph of $f(x) = 8^x$ has y-coordinate the square of the y-coordinate of the point vertically below it on the graph of $8^{0.5x}$.

c) Under a reflection in the y-axis, "$-x$" replaces "x" and conversely, so, the graph of $f(x) = 8^{Ax}$ is mapped onto the graph of $f(x) = 8^{-Ax}$.

5. a) The condition the $f(x+d) = 2f(x)$ where $f(x) = b^{Ax}$ can be written as $b^{Ax+Ad} = 2b^{Ax}$. That is, $b^{Ad}b^{Ax} = 2b^{Ax}$.
Dividing both sides of this equation by b^{Ax}, we obtain $b^{Ad} = 2$.
Since $A > 0$ and $b > 1$, then the graph of $f(x) = b^{Ax}$ is similar to the graph of $f(x) = 8^{Ax}$ shown above, i.e. the values of $f(x)$ increase from 1 through all the real numbers including 2.
Therefore, there is a value of x which satisfies the equation $b^{Ax} = 2$. This value of x is the doubling time, d.

b) The required equation is $b^{Ad} = 2$.

c) We graph $Y_1 = 1.16^x$ and $Y_2 = 2$ with window settings: $0 \le x \le 10$; $0 \le y \le 3$. Then select 5 (intersect) from the [CALC] menu to obtain the display shown. We observe that Y_1 has doubled its original value when $x = 4.67$ years or 4 years and 8 months.

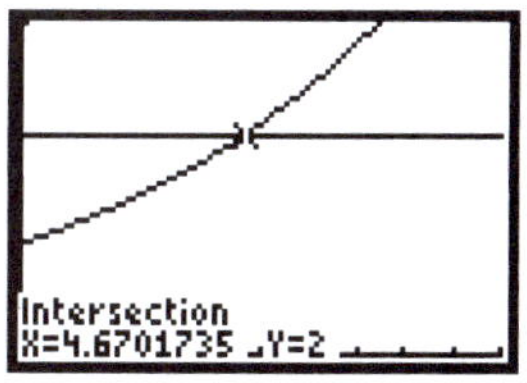

6. a) Proceeding as in exercise 5, we write the condition $f(x+h) = 0.5f(x)$ as $b^{At+h} = 0.5b^{At}$. This reduces to $b^{Ah} = 0.5$.
Since b^{At} decreases ($A < 0$) from 1 toward 0 through all real numbers including 0.5, it assumes the value 0.5 for some value of t which we denote, h. This is the half-life.

b) The required equation is $b^{Ah} = 0.5$.

c) We graph $Y_1 = e^{-0.0004332x}$ and $Y_2 = 0.5$ in the window $0 \le x \le 4000$; $0 \le y \le 1$. Then we select 5 (intersect) from the [CALC] menu to obtain the display shown here. The intersection point is (1600…, .5). That is, the half-life of the isotope is about 1600 years.

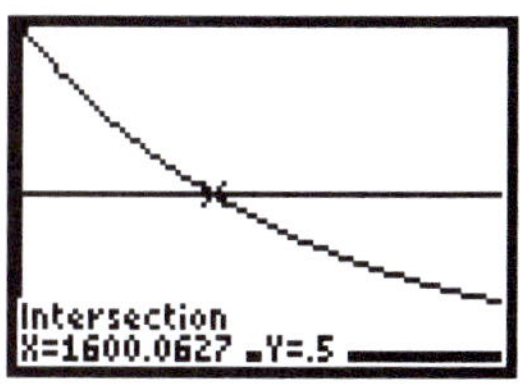

Exploration 15

1. a) log 1000 = 3 b) log 100 000 = 5 c) log 0.001 = −3

2. a) $y = \log x$ means $x = 10^y$.
b) $\log x = 4$ means $x = 10^4$ or 10 000.

3. Expressed to 5 decimal places:
a) 0.30103 b) 1.30103 c) 2.30103 d) 3.30103

4. Multiplication by 10, 100 or 1000 causes the logarithm to increase by 1, 2 or 3 respectively.

5. a) By definition, $10^{\log 2} = 2$ and $10^{\log 20} = 20$, therefore $10^a = 2$ and $10^b = 20$. That is, $10^b = 10 \times 10^a$. Using the exponent law for multiplying powers of 10, we have $10^b = 10^{a+1}$ so, $b = a + 1$.
b) The derivation in part a verifies the observation in exercise 4 that multiplication by 10 increases the logarithm of a number by 1.
c) The graph of $\log 10x$ is the graph of $\log x$ translated one unit vertically.

6. a) When we graph $Y_1 = \log x$, $Y_2 = \log 10x$ and $Y_3 = \log 100x$, in the window $0 < x \le 100$; $-1 \le y \le 5$, we obtain the display on the right.

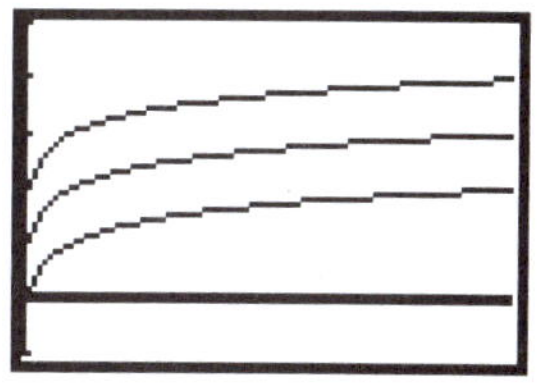

b) When we trace along the curve $y = \log x$, to the point (50, 1.69897), we press the "up cursor" key and the cursor moves to the point (50, 3.69897) on the graph of $y = \log 100x$. In general, if (x, y) is an arbitrary point on the graph, of $\log x$, then the points on the graphs of $\log 10x$ and $\log 100x$ directly above (x, y) are $(x, y + 1)$ and $(x, y + 2)$ respectively.

7. a) Since $\log x = 2.4$, then $10^{\log x} = 10^{2.4}$. We compute $10^{2.4}$ directly, as 10^2.4 and obtain 251.1886432… That is, $10^{2.4} = 251.1886432\ldots$
Since $10^{\log x} = x$, it follows that $x = 251.188\ldots$

b) From part a), log 251.1886432 = 2.4, therefore, log 2.51188… = 0.4. Therefore, 2log2.51188… = 0.8. That is, $\log 2.51188^2 = 0.8$. Squaring yields log 6.3095… = 0.8, and so, from *exercise* 6, we conclude, log 6309.5… = 3.8; that is, $x = 6309.5$. Alternatively, we proceed as in part a) and compute x = 10^3.8 which is 6309.573…

c) In part b), we found log 6.3095… = 0.8, so $x = 6.3095$.

d) We can either compute x directly as $10^{-2.4}$, or compute it as the reciprocal of $10^{2.4}$ which we found in part a) to be 251.188… In either case, we find, $x = 0.00398\ldots$

8. Applying the multiplicative property of logarithms, we write: $\log x + \log x^{-1} = \log x x^{-1}$ which yields log 1 which is 0. That is, the logarithm of a number and its reciprocal add to zero.

9. a) The graphs are shown in the display.
b) The graph of Y_3 is the horizontal line.
c) $Y_2 - Y_1$ is the line y = 0.69897…
d) In part c), we observe that 0.69897 is log 5, so the graph of $\log 5x$ is the graph of $\log x$ translated log 5 units vertically. We conjecture that the graph of $\log Ax$ is the graph of $\log x$ translated vertically

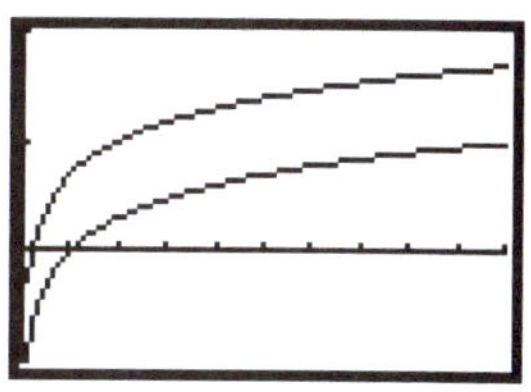

Exploration 15 (cont'd)

10. a) When we graph $Y_1 = \log x + \log 5$, and $Y_2 = \log 5x$, we get the single graph shown in the display below left.
b) When we graph $Y_3 = Y_2/Y_1$, we get the display shown below right.

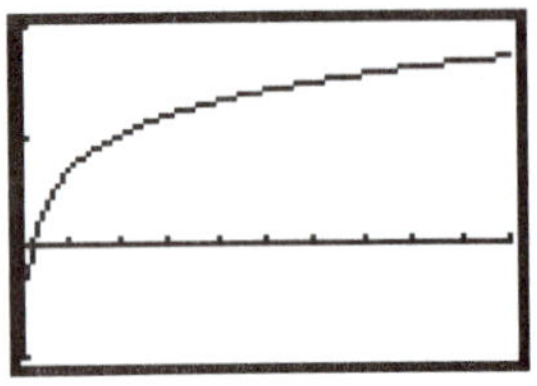

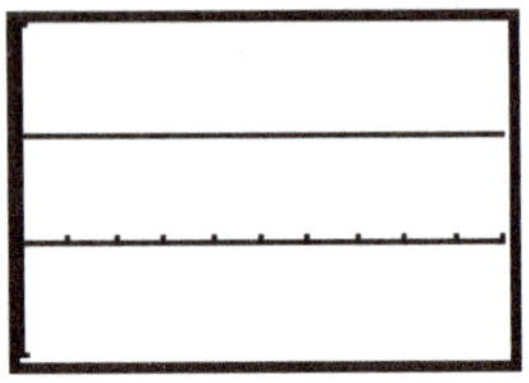

c) All points on the graph of Y_3 have second coordinate equal to 1.
d) The results in part c) lead us to the same conjecture as in exercise 9, that is; $\log Ax = \log x + \log A$.

11. (i) By definition, $10^{\log a} = a$ and $10^{\log b} = b$ and $10^{\log ab} = ab$, so:

$$ab = 10^{\log a} \bullet 10^{\log b}$$
$$= 10^{\log a + \log b} \quad \text{(exponent laws)}$$
$$\text{But } ab = 10^{\log ab} \quad \text{(by definition)}$$
$$\text{So, } \log ab = \log a + \log b$$

(ii) Substituting b^{-1} for b, in part (i) yields:

$$\log ab^{-1} = \log a + \log b^{-1}$$
$$= \log a - \log b \quad \text{(See exercise 8)}$$

12. a) Using the LOG key, we find $\log 4 = 0.60206\ldots$
b) Expressed to 5 decimal digits:
(i) $\log 40 = \log 10 + \log 4 = 1.60206$
(ii) $\log 400 = \log 100 + \log 4 = 2.60206$
(iii) $\log 4000 = \log 1000 + \log 4 = 3.60206$
(iv) $\log 1/4 = \log 1 - \log 4 = -0.60206$
(v) $\log 1/40 = -\log 40 = -1.60206$
(vi) $\log 1/400 = -\log 400 = -2.60206$
(vii) $\log 2.5 = \log 10/4 = \log 10 - \log 4 = 0.39794$
(viii) $\log 25 = 1.39794$
(ix) $\log 250 = 2.39794$

Exploration 16

1. The asymptotic behavior of a function refers to its limiting values as $x \to -\infty$ and/or as $x \to \infty$.

2. Graphing the functions and tracing along the curves, we obtain the following asymptotic limits.
a) $f(x) \to \frac{14}{3}$ as $x \to -\infty$ and as $x \to \infty$.
b) $F(x) \to 0$ as $x \to -\infty$ and as $x \to \infty$.
c) $g(x) \to 0$ as $x \to -\infty$ and $g(x) \to 8.8$ as $x \to \infty$.
d) $G(x) \to 7$ as $x \to -\infty$ and as $x \to \infty$.

3. We obtain the following graphs from which we obtain the asymptotic limits shown below each display.

a)

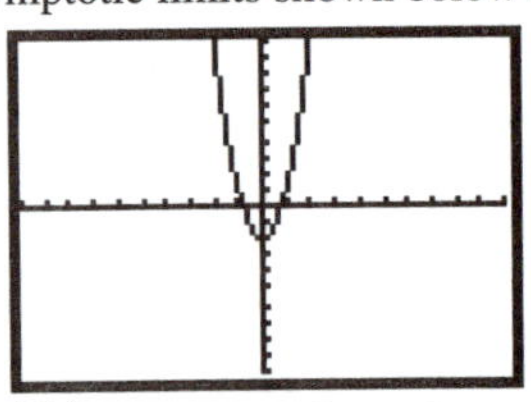

$f(x) \to \infty$ as $x \to \infty$ or $-\infty$.

b)

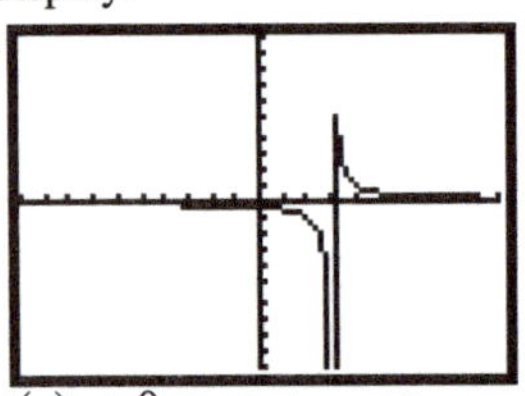

$g(x) \to 0$ as $x \to \infty$ or $-\infty$.

Exploration 16 (cont'd)

3. c)

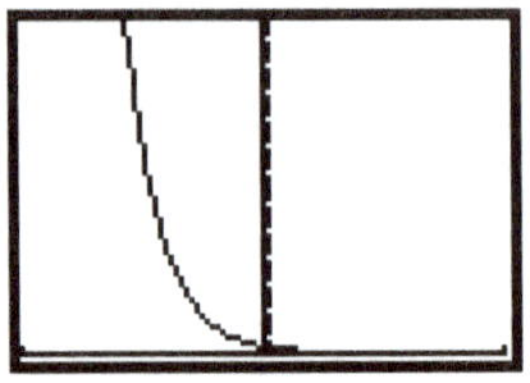

using the window: $-1 \le x \le 1$
$0 \le y \le 50$
$h(x) \to \infty$ as $x \to -\infty$
$h(x) \to 0$ as $x \to \infty$

d)

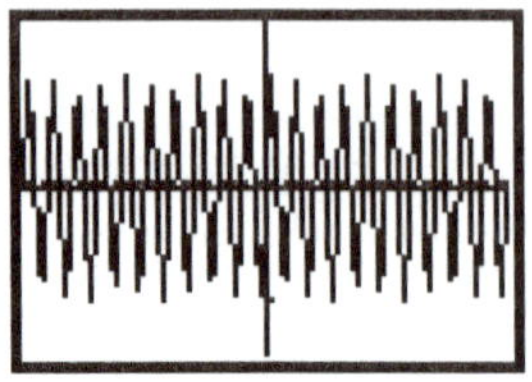

window $-20\pi \le x \le 20\pi$
$-1.5 \le y \le 1.5$
$m(x)$ varies between -1 and $+1$ and has no asymptotic limit as $x \to -\infty$ or ∞

4. a) no unbounded limit exists b) 0 c) 0 d) no limit

5. $f\left(\frac{1}{x}\right) = \left(\frac{3}{x^2}\right) - 2$ $\quad g\left(\frac{1}{x}\right) = \frac{x}{1-3x}$ $\quad h\left(\frac{1}{x}\right) = 10^{-\frac{3}{x}}$

6. a) ∞ b) 0 c) 0

7. The results in exercises 4 and 6 are the same because

$$\lim_{x\to-\infty} f(x) = \lim_{x\to0^-} f\left(\frac{1}{x}\right) \quad \text{and} \quad \lim_{x\to\infty} f(x) = \lim_{x\to0^+} f\left(\frac{1}{x}\right)$$

8. a) 3/2 b) 0 c) 0 d) 9

9. a) When we graph $Y_1 = \frac{\sin x}{x}$ we obtain this display.

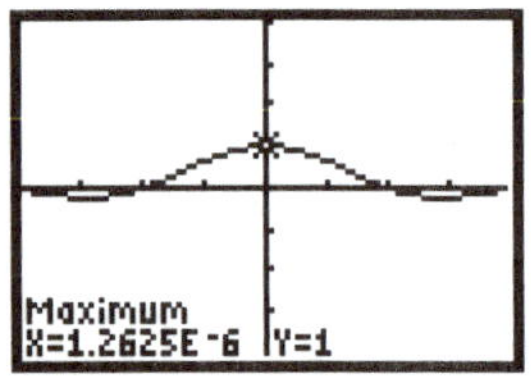

b) As we trace along this curve, we observe that at $x = -.1308997$, $y = 0.99714666$. At $x = .13089969$, $y = 0.99714666$. It appears that

$$\lim_{x\to0} \frac{\sin x}{x} = 1$$

To check this, we select 4 (maximum) from the [CALC] menu and find the "maximum" point (1.2625E–6, 1). That is, as $x \to 0$, $\frac{\sin x}{x} \to 1$

10. a) $\lim_{x\to0} \frac{\sin 2x}{x} = 2 \lim_{x\to0} \frac{\sin 2x}{2x} = 2$

b) $\lim_{x\to0} \frac{\sin x}{2x} = \frac{1}{2} \lim_{x\to0} \frac{\sin x}{x} = \frac{1}{2}$

c) $\lim_{x\to0} \frac{\tan 3x}{x} = 3 \lim_{x\to0} \frac{\sin 3x}{3x} = 3$

d) $\lim_{x\to0} \frac{5x}{\sin 2x} = \frac{5}{2} \lim_{x\to0} \frac{x}{\sin x} = \frac{5}{2}$

11. Substitute $\frac{1}{x}$ for x: $\lim_{x\to\infty} \frac{(6x+1)^4 - (6x)^4}{x^3} = \lim_{x\to0} \frac{(6+x)^4 - 6^4}{x}$

We can factor the numerator on the right side using a difference of squares. As $x \to 0$, the right side approaches 864. To verify this answer, we could graph the function, $f(x)$ where $f(x) = \frac{(6+x)^4 - 6^4}{x}$

Answers to Exercises & Hints for Investigations

Exploration 16 (cont'd)

11. (cont'd) We graph $f(x)=\frac{(6+x)^4-6^4}{x}$ with window settings:

$-10 \le x \le 10;\ 0 \le y \le 2000$ and we obtain the graph shown in the display. We trace along the curve to the point (0.21276596, 911.05354). Then we zoom in and trace along the curve to the point (3E–9, 864). This verifies that

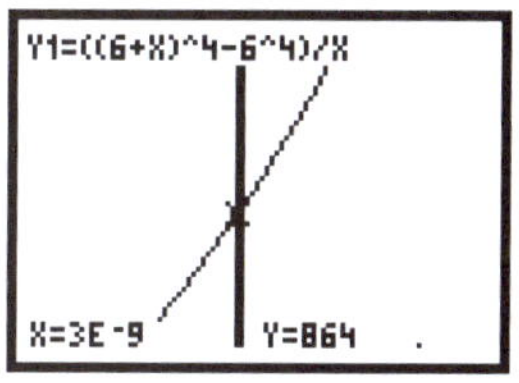

$$\lim_{x\to 0}\frac{(6+x)^4-6^4}{x}=864$$

12. $F\left(\frac{1}{x}\right)=\left(\frac{1}{x}\right)^2[1-\cos x]$. When we graph $F\left(\frac{1}{x}\right)$, with window settings:
$-10 \le x \le 10$; $-1 \le y \le 1$, we obtain the graph shown in the display.

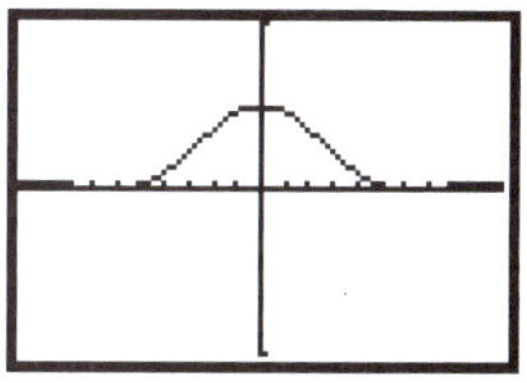

As we trace along the curve, close to the y-axis, we reach the points, (-0.212766, 0.49811662) and (0.212766, 0.49811662). To determine the coordinates of the maximum point, we select 4 (maximum) on the [CALC] menu and we obtain the point (–1.073E–4, 0.50000041..). That is, $\lim_{x\to\infty}F(x)=0.5$.

13. We define $Y_1=\frac{\sqrt{30x+30}-\sqrt{30}}{x}$

On graphing Y_1, we obtain this display. To determine where this curve intersects the y-axis, we trace along the curve to $x = 0$ and zoom in. We reach the point (3E-9, 2.7386). This suggests that

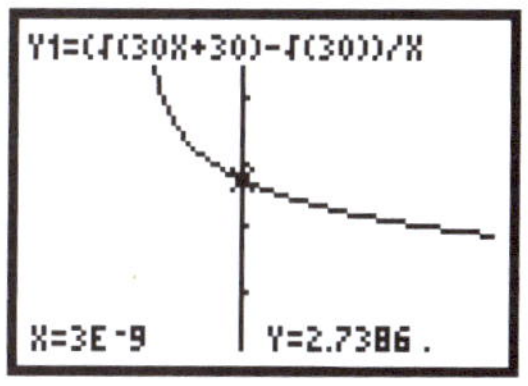

$$\lim_{x\to 0}\frac{\sqrt{30x+30}-\sqrt{30}}{x}=2.74 \text{ (to 2 decimal places)}$$

By rationalizing the numerator, and evaluating the limits of the result, we can verify that the limit is $\sqrt{30}/2$ or 2.7386...

Hints for the Investigations:

To evaluate $\lim_{x\to\infty}\left(1+\frac{1}{x}\right)^{cx}$, you may assume that,

$$\lim_{x\to\infty}\left(1+\frac{1}{x}\right)^{cx}=\left(\lim_{x\to\infty}\left(1+\frac{1}{x}\right)^{x}\right)^{c}$$

Similarly, to evaluate $\lim_{x\to\infty}\left(1+\frac{A}{x}\right)^{cx}$, we can substitute $y=\frac{x}{A}$ and write,

$$\lim_{x\to\infty}\left(1+\frac{A}{x}\right)^{cx}=\left(\lim_{y\to\infty}\left(1+\frac{1}{y}\right)^{y}\right)^{cA}$$

In *exercise* 18, use the identity $\lim_{x\to\infty}f(x)=\lim_{x\to 0^+}f\left(\frac{1}{x}\right)$

Exploration 17

1. a) $v_{av}=\Delta s/\Delta t$

b) To determine the average velocity between time t_0 and t_1, locate the points on the distance-time graph corresponding to these times. Calculate the slope of the secant joining these two points.

2. a) $a_{av}=\Delta v/\Delta t$

b) To determine the average acceleration between time t_0 and t_1, locate the points on the velocity-time graph corresponding to these times. Calculate the slope of the secant joining these two points.

c) Since the velocity is a linear function of time, the average velocity is the arithmetic mean of the velocities at $t = 2.8$ and $t = 9.2$, which is $9.8[2.8 + 9.2]/2 = 58.8$ *m/s*.

3. Yes. The development in *worked example* 1 shows that if the acceleration is a constant a, then $v = v_0 + at$ where v_0 is the initial velocity. That is, the velocity is a linear function of time.

4. a) The cannon ball reaches the ground when $4.9t^2 = 56$; i.e. when $t = \sqrt{(56/4.9)} \approx 3.38$, i.e. it hits the ground in about 3.38 *s*.
b) No. The cannon ball is accelerating, so it travels farther in the second half of the time interval. The distance traveled after 3.5 *s* would be less than half the distance traveled in that one-second interval.

5. a) The **Value** command for $t = 10$ yields the distance 102.923, as shown on the display, indicating that Johnson traveled about 103 meters in the first 10 seconds. Similarly, we find that he traveled about 168.5 meters in the first 15 seconds.

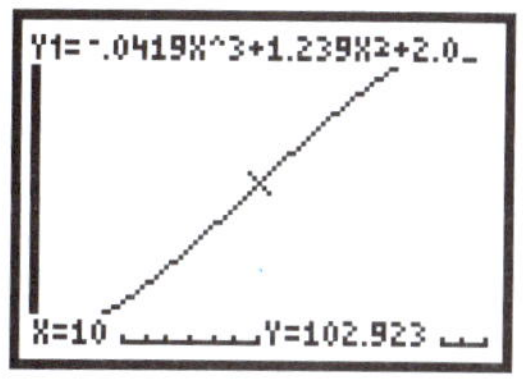

b) His average velocity in the interval $10 \le x \le 15$ was about (168.5 – 103)/5 or 13.1 meters per second.

c) His instantaneous velocity at $t = 10$ was $dy/dx \approx 14.3$ m/s.

d) His "actual" velocity at the 10 second mark was greater than his average velocity over the next 5 seconds.

6. a) $s + \Delta s = 4.9(t + \Delta t)^2$.
b) $\Delta s = 4.9(t + \Delta t)^2 - 4.9t = 9.8t(\Delta t) + (\Delta t)^2$.
c) $\Delta s/\Delta t = 9.8t + \Delta t$
d) $9.8t$

Exploration 18

1. a) Solving for y, we obtain: $y=\pm\frac{b}{a}\sqrt{a^2-x^2}$
This indicates that there are two values of y corresponding to each value of x, when $x < a$. This violates the condition that there is a unique y for each x. Therefore it is not a function.

b) A vertical line $x = c$ where $-a < c < a$, intersects the graph in two points indicating that there are two points on the graph with the same x-coordinate and different y-coordinates.

2. a) $\frac{x^2}{9^2}+\frac{y^2}{6^2}=1$ b) 18 c) $y \approx \pm 3.77$ when $x = 7$

Answers to Exercises & Hints for Investigations

Exploration 18 (cont'd)

3. The equation is $\frac{x^2}{10^2} + \frac{y^2}{8^2} = 1$

4. a) (i) $(\pm5, 0)$ and $(0, \pm4)$ (ii) $(\pm3, 0)$

b) Solving for y yields, $y = \pm\frac{4}{5}\sqrt{25-x^2}$

To graph this curve, we define $Y_1 = \frac{4}{5}\sqrt{25-x^2}$ and $Y_2 = -Y_1$.

When we graph these two functions, using the Zoom 5 window settings, we obtain the ellipse shown in the display on the right. Observe that as we trace along this "ellipse", we are really tracing along the graphs of two distinct functions. We must press the "cursor up" or "cursor down" key to move from one curve to the other. Since the two curves are not joined, we cannot reach the x-intercepts by tracing. *This problem does not occur when the ellipse is defined using a polar equation.*

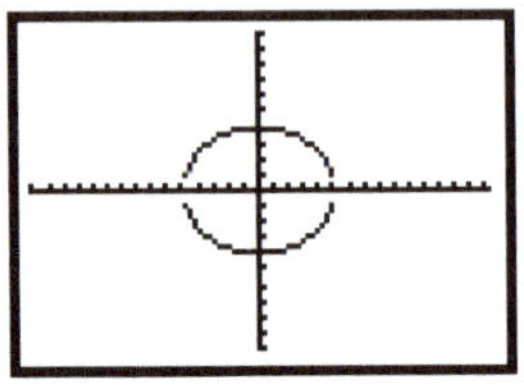

5. a) To graph this curve, we solve for y as in exercise 4 and define $Y_1 = \frac{5}{4}\sqrt{16-x^2}$ and $Y_2 = -Y_1$.
When we graph Y_1 and Y_2 using the Zoom 5 window settings, we obtain the display shown here. The intersections with the coordinate axes found algebraically, are $(\pm4, 0)$ and $(0, \pm5)$.

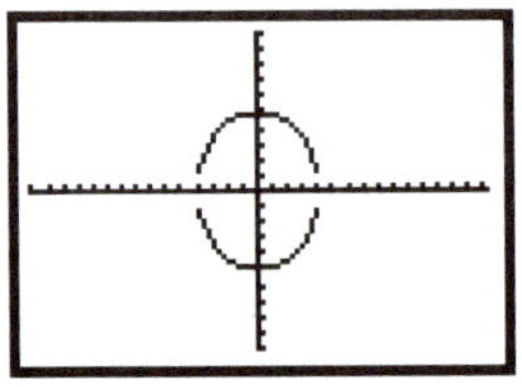

b) Any point on the line $y = x$ will remain unchanged under a reflection in that line. Therefore we can find intersection points by solving for the intersections of the ellipse with the line $y = x$. Similarly, we can solve for intersections with the line $y = -x$.

6. Multiplication of equation ① by 4 and subtracting equation ② yields $187y^2 = 2560$. Therefore, $y = \pm3.6999\ldots$ from which we calculate $x = 2.6598\ldots$ These are the same as the values obtained using the **intersect** command on the TI-84 Plus.

7. a) The completed table is shown below.

	R	P	R^3	P^2	R^3/P^2
Mercury	0.39	0.24	0.05932	0.0576	1.0298
Venus	0.72	0.62	0.37325	0.3844	0.9710
Mars	1.52	1.88	3.5118	3.5344	0.9936
Jupiter	5.20	11.86	140.608	140.659	0.9996
Saturn	9.55	29.46	870.983	867.891	1.0035

The numbers in the last column are close to 1, verifying Kepler's third law within the limits of measurement precision. Also the rounding of these numbers results in deviations from 1 because small discrepancies in the denominator lead to much larger discrepancies in the quotient R^3/P^2.

b) $P^2 = R^3$, so $P = R^{3/2}$. Therefore Neptune's orbital period is $30.1^{3/2}$ or about 165 earth years.

Exploration 18 (cont'd)

8. a) To graph this orbit, we follow the procedure in the *worked example* 3 using the equation below and we obtain the graph shown in the display.

$$\frac{x^2}{1.523^2} + \frac{y^2}{1.5164^2} = 1$$

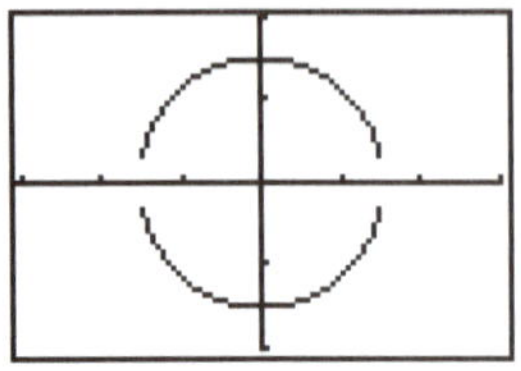

b) When we trace around the curve, we discover that after the cursor reaches the point $(1.48\ldots, 0.3415\ldots)$, it disappears from the graph. This is a consequence of the fact that the two branches of this ellipse are not joined. To trace around an ellipse, we must graph it in a different mode called *polar mode* that we will introduce in *exploration 20*.

9. a) An ellipse with $e = 0$ has $a = b$ and so it is a circle.

b) Since $c^2 = a^2 - b^2$, then for the orbit of Mars, $c = 0.1416\ldots$ Then $e = c/a$ or 0.093 for the planet Mars.

10. a) The condition that P moves so that its distance from F is e times its distance from the line with equation $x = d$ can be written as:

$$\sqrt{(x-c)^2 + y^2} = e(x - d)$$

Squaring and simplifying yields the equation:

$$(1 - e^2)x^2 + 2x(e^2d - c) + y^2 + (c^2 - e^2d^2) = 0.$$

b) If the ellipse is centered at the origin, then the coefficient of the x term is 0, i.e., $e^2d = c$, so $d = c/e^2$.

c) The equation becomes: $\frac{x^2}{(c/e)^2} + \frac{y^2}{(c/e)^2(1-e^2)} = 1$

d) $a = c/e$ and $b = (c/e)\sqrt{1-e^2}$ e is the same eccentricity.

Exploration 19

1. See answers to exercise 1 of *Exploration* 18.

2. a) $\frac{x^2}{4^2} - \frac{y^2}{6^2} = 1$ b) 8 c) There are no points on the hyperbola such that $-4 \le x \le 4$.

4. a) i) $(\pm5, 0)$ ii) $(\pm\sqrt{41}, 0)$

b) To graph this curve, solve for y and define $Y_1 = \frac{4}{5}\sqrt{x^2 - 25}$ and $Y_2 = -Y_1$.
When we graph Y_1 and Y_2 using the default window settings, we obtain the display shown here. We observe that the graph is a hyperbola which does not intersect the y-axis. Applying the **intersect** command verifies the branches intersect at $(\pm5, 0)$.

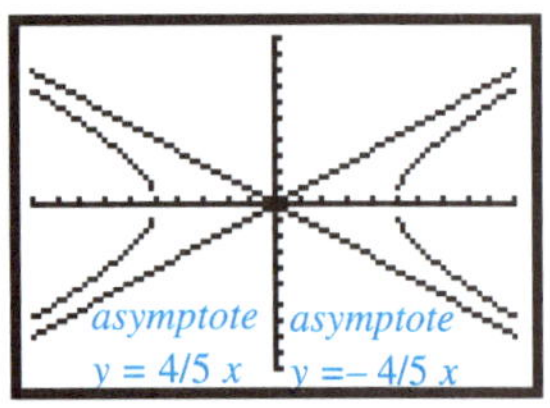

7. b) The display shows the four intersection points of the hyperbolas with equations: $\frac{(x-7)^2}{9} - \frac{y^2}{40} = 1$

and $\frac{x^2}{12} - \frac{(y+4)^2}{4} = -1$

The intersection point is shown here.

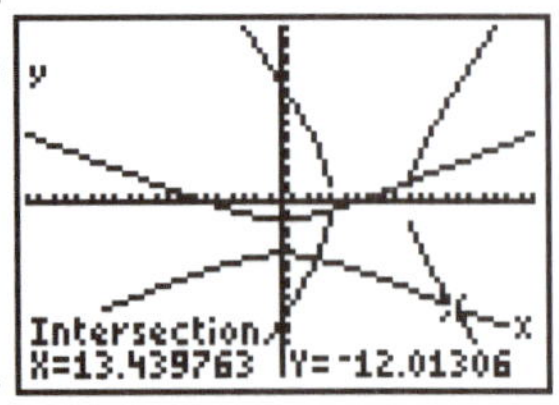

Exploration 20

1. The equations relating polar coordinates to Cartesian coordinates are:

$$x = r\cos\theta$$
$$y = r\sin\theta$$

Substituting $r = 5$, $\theta = \pi/2$ into these equations yields the Cartesian coordinates (0, 5). Point P is 5 m above the origin and the polar axis.

2. a) Applying the Theorem of Pythagoras yields the values $r = 5$, 13, $7\sqrt{2}$, and 2 for points P, Q, R, and S respectively. Evaluating $\tan^{-1}(y/x)$ yields 53.13°, 112.62°, 315°, and 240° for points P, Q, R, and S respectively.

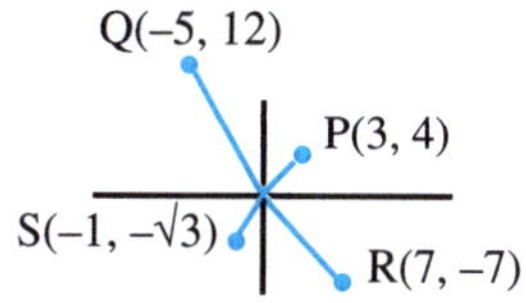

These answers are not unique because the inverse tangent is not a function. e.g. P(3, 4) can be written in polar coordinates as (5, 53.13°) or (–5, 233.13°).

b) The equation expressing r in terms of x and y is $r = \pm\sqrt{x^2+y^2}$.

c) The equation expressing θ in terms of x and y is $\theta = \tan^{-1}\left(\frac{y}{x}\right)$.

3. a) Using the equations (given in exercise 1) to convert these polar coordinates into Cartesian coordinates, we obtain:

$$A\left(\frac{1}{\sqrt{2}}, \frac{1}{\sqrt{2}}\right), B(0, 5), C(-2, 2\sqrt{3}), \text{and } D\left(-\frac{\sqrt{3}}{2}, -\frac{1}{2}\right)$$

b) The formula expressing x in terms of r and θ is: $x = r\cos\theta$.

c) The formula expressing y in terms of r and θ is: $y = r\sin\theta$.

4. a) The points of intersection with the polar axis have coordinates ($n\pi$, $n\pi$) where n is an integer. These points of intersection are distributed a distance of 2π units apart.

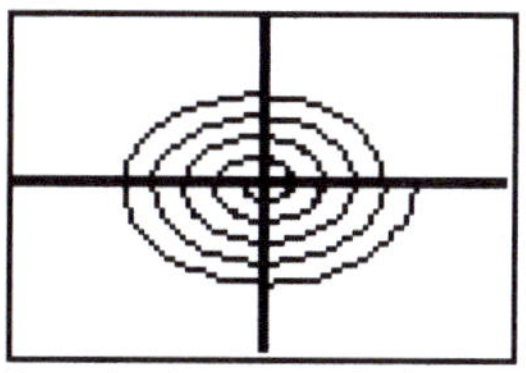

b) The number of intersections with the polar axis is the number of multiples of π between θmin and θmax.

5. The points of intersection are given below each graph.

a)

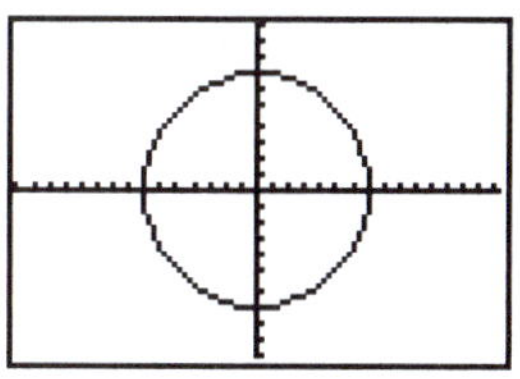

(7, 0) and (7, π)

b)

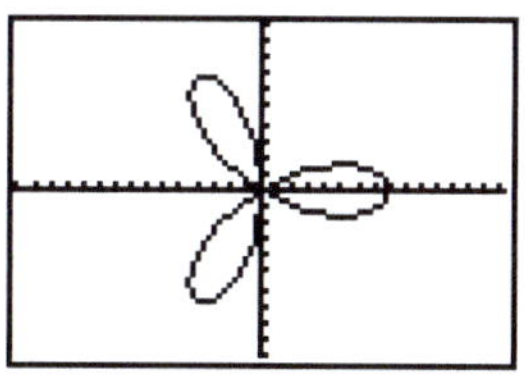

(0, $(2n-1)\pi/6$) and (8, $n\pi$)

The polar graph intersects the polar axis are those points (r, θ) for which $r = 0$ and those points for which $\theta = n\pi$ where n is a non-negative integer. We determine these points by setting $r = 0$ and solving the resulting equation, and setting $\theta = n\pi$ to calculate r.

c)

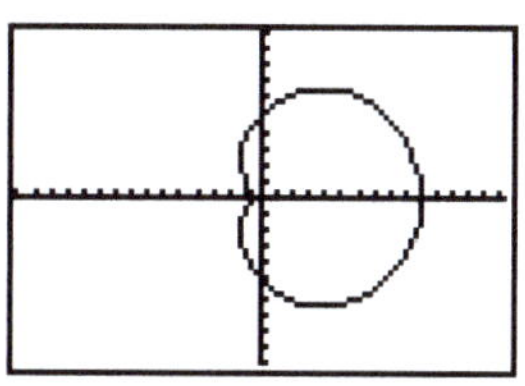

(0, $(2n-1)\pi$) and (10, $2n\pi$)

6. a)

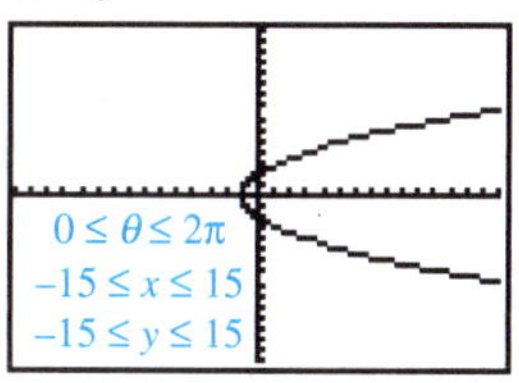

b)

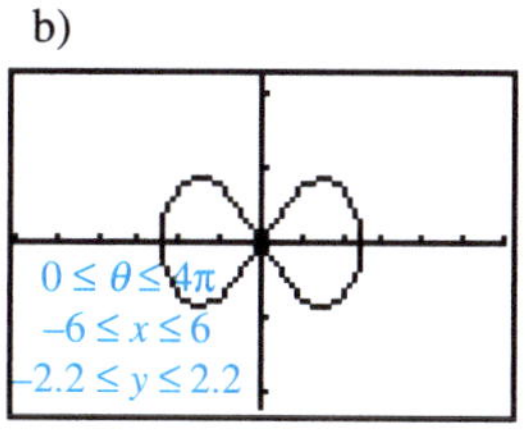

Exploration 20 (cont'd)

7. Substituting the values of a and b into the polar equation for an ellipse yields the following equation of the orbit of Mars.

$$r = \frac{1.5164^2}{1.523 - \sqrt{1.523^2 - 1.5164^2}\cos\theta}$$

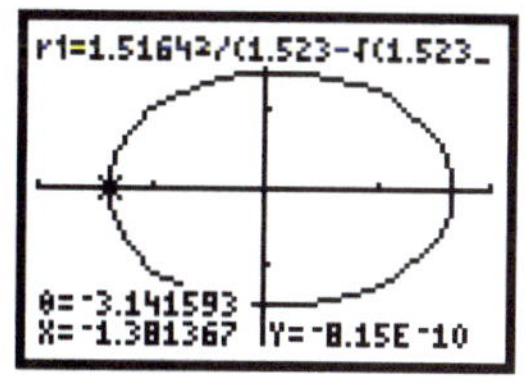

We graph r_1 in polar mode using the window $-2 \le x \le 2$; $-2 \le y \le 2$, and we obtain the graph in the display. When we trace along this curve, we find the aphelion at $x = 1.6646335$ and the perihelion at $x = -1.381367$.

Exploration 21

1. a) $2x - y + 1 = 0$ b) $x^2 + y^2 = 4$

2. a) We substitute $T = \frac{x}{49}$ into the equation for y to obtain:

$$y = 20\left(\frac{x}{49}\right) - 4.9\left(\frac{x}{49}\right)^2 \text{ so, } 490y = 200x - x^2.$$

b) From the equation in part a), we see that $y = 0$ when $x = 200$, so we graph the parametric equations in the window $0 \le x \le 200$; $0 \le y \le 40$; $0 \le \text{T} \le 2\pi$; Tstep = .1, to obtain a parabola as in the Worked Example. Upon tracing, we reach the point X = 98, Y= 20.4 and T = 2 i.e., the golf ball reaches its maximum height of about 20.4 m approximately 2 seconds after impact.

c) The golf ball is in the air about 2 × 2 or 4 seconds.

3. a) When we graph these equations using the default window, we obtain the display shown here.

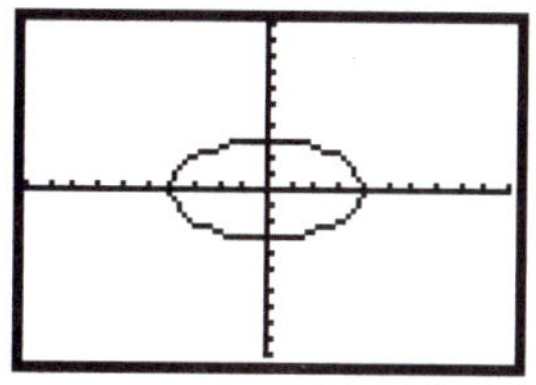

b) This curve is an ellipse.

c) $\left(\frac{x}{4}\right)^2 + \left(\frac{y}{3}\right)^2 = 1$

d) The equation shows that the intersections with the coordinate axes occur at (±4, 0) and (0, ±3).

4. From the innermost to the outermost ellipse, the equations are:

$X_{1T} = 4\cos T$, $Y_{1T} = 6\sin T$ $X_{2T} = 6\cos T$, $Y_{2T} = 3\sin T$ $X_{3T} = 6\cos T$, $Y_{3T} = 6\sin T$

5. When we graph these parametric equations in the range $0 \le \text{T} \le 4\pi$, we obtain the graph shown in the display.

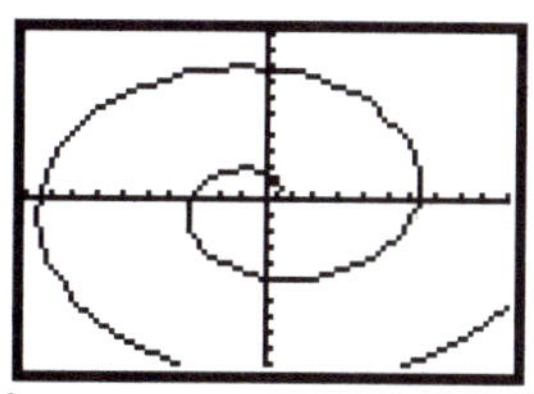

a) This is the *spiral of Archimedes.*

b) We can express x and y in terms of T by drawing a right triangle with hypotenuse T units in length and one angle having a measure of T radians. Then,

$$T = \tan^{-1}\left(\frac{y}{x}\right) \text{ and } x^2 + y^2 = T^2$$

$$\text{i.e., } x^2 + y^2 = \tan^{-1}\left(\frac{y}{x}\right)^2$$

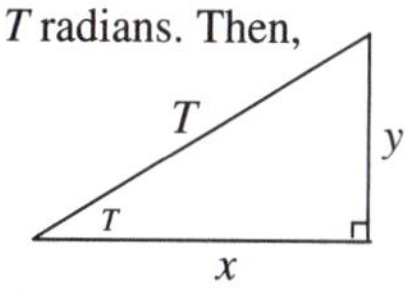

This is a single equation in x and y but it cannot be solved for y in terms of x and therefore you cannot graph it in Cartesian form on your TI-84 Plus. However, you can graph it in parametric form as in part a), or in polar form using the equation, $r = \theta$.

TIMESAVER TI-89 TRILOGY for MATHEMATICS TEACHERS

•Complete ready-made instructional lessons

•Step-by-Step Keying Sequences

•Historical Anecdotes & Humor

•Detailed Solutions

$16.95 96 pages

20 Explorations develop the fundamental concepts of differential & integral calculus using the TI-89.

Topics included are:

- sequences & series
- limits
- derivatives of rational & transcendental functions
- related rates & motion
- equations of tangents to curves
- difference equations
- integration
- techniques of integration
- arc length
- surface area and volumes of solids of revolution
- differential equations
- 3-D Plots

$16.95 96 pages

20 Explorations develop the concepts of advanced algebra using the TI-89.

Topics included are:

- Fundamental Theorem of Algebra
- polynomial functions
- polynomial equations
- rational functions
- trigonometric functions
- exponential functions
- logarithmic functions
- linear systems
- linear programming
- matrices
- congruence arithmetic
- complex numbers
- fractals

$16.95 96 pages

20 Explorations develop the concepts of probability & statistics using the TI-89.

Topics included are:

- statistical inference
- box plots, histograms
- central tendency
- measures of dispersion
- scatterplots & correlation
- sampling
- regression & line of best fit
- combinatorics
- probability
- Monte Carlo Techniques
- binomial distribution
- normal distribution

ORDER FORM

Get The Full Set!

Free! *2-CDPack*

With every trilogy we will send a free copy of the sensational 2-CD pack *The Psychology of Empowerment* by Dr. Gene Landrum (See back page)

To order, please indicate quantities in the table below and calculate total payment.

ISBN	QTY	TITLE	UNIT PRICE	NET PRICE
1-895997-25-9		FUNCTIONS WITH THE TI-84 PLUS CALCULATOR	$19.95	
1-895997-27-5		STATISTICS WITH THE TI-84 PLUS CALCULATOR	$19.95	
1-895997-26-7		ALGEBRA WITH THE TI-84 PLUS CALCULATOR	$19.95	
		The TI-84 Plus Trilogy (Set of all three TI-84+ titles above)	**$50**	
1-895997-04-6		A MATHEMATICAL MOSAIC: PATTERNS & PROBLEM SOLVING	$16.95	
1-895997-12-7		ADVANCED ALGEBRA WITH THE TI-89	$16.95	
1-895997-14-3		STATISTICS & PROBABILITY WITH THE TI-89	$16.95	
1-895997-13-5		CALCULUS WITH THE TI-89	$16.95	
		The TI-89 Trilogy (Set of all three TI-89 titles above)	**$45**	
POSTAGE AND HANDLING				$5.00

Fax this order form to (905) 335-5104.
OR
Make check payable to Brendan Kelly Publishing Inc. and forward with this form to the address here. →
OR
Order on line at: www.brendankellypublishing.com

TOTAL $US ________

Brendan Kelly Publishing Inc.
2122 Highview Drive
Burlington, Ontario
CANADA L7R 3X4
Telephone: (905) 335-3359

Ship to:

Name: ______________________________

Address: ______________________________

City: ______________________________

State: ________ Zip: ________________